Understanding Space Strategy

This book examines the rise of great power competition in space, including the relevant and practical space strategies for China, Russia, the United States, and other countries.

The work discusses the concepts and writings of past strategists, such as Thucydides, Sun Tzu, and Clausewitz, in relation to warfare initiated in or extending into space. This analysis underscores why polities initiate war based upon an assessment of fear, honor, and interest, and explains why this will also be true of war in space. Based upon the timeless strategic writings of the past, the book uncovers the strategy of space warfare, along with the concepts of deterrence, dissuasion, and the inherent right of self-defense, and outlines strategies for great, medium, and emerging space powers. Additionally, it highlights changes needed to space strategy based upon the Law of Armed Conflict, norms of behavior, and Rules of Engagement. The work also examines advancements and emerging trends in the commercial space sector, as well as what these changes mean for the implementation of a practical space strategy. Given the rise of great power competition in space, this work presents a space strategy based upon historical experience.

This book will be of much interest to students of space policy, strategic studies, and International Relations.

John J. Klein is a Senior Fellow and Strategist at Falcon Research, USA; and serves as an adjunct professor at the George Washington University's Space Policy Institute. He is the author of *Space Warfare: Strategy, Principles and Policy* (2006).

Space Power and Politics
Everett C. Dolman and John Sheldon
School of Advanced Air and Space Studies, USAF Air, Maxwell, USA

The Space Power and Politics series will provide a forum where space policy and historical issues can be explored and examined in-depth. The series will produce works that examine civil, commercial, and military uses of space and their implications for international politics, strategy, and political economy. This will include works on government and private space programs, technological developments, conflict and cooperation, security issues, and history.

Chinese Space Policy
A study in domestic and international politics
Roger Handberg and Zhen Li

The International Politics of Space
Michael Sheehan

Space and Defense Policy
Edited by Damon Coletta and Frances T. Pilch

Space Policy in Developing Countries
The search for security and development on the final frontier
Robert C. Harding

Space Strategy in the 21st Century
Theory and policy
Edited by Eligar Sadeh

Transatlantic Space Politics
Competition and cooperation above the clouds
Sheng-Chih Wang

Understanding Space Strategy
The Art of War in Space
John J. Klein

For more information about this series, please visit: www.routledge.com/strategicstudies/series/SPP

Understanding Space Strategy
The Art of War in Space

John J. Klein

Routledge
Taylor & Francis Group

LONDON AND NEW YORK

First published 2019
by Routledge
2 Park Square, Milton Park, Abingdon, Oxon OX14 4RN

and by Routledge
52 Vanderbilt Avenue, New York, NY 10017

First issued in paperback 2020

Routledge is an imprint of the Taylor & Francis Group, an informa business

British Library Cataloguing-in-Publication Data
A catalogue record for this book is available from the British Library

Library of Congress Cataloging-in-Publication Data
Names: Klein, John J., author.
Title: Understanding space strategy : the art of war in space / John J.
Klein.
Description: First edition. | Milton Park, Abingdon, Oxon ; New York,
NY : Routledge, [2019] | Series: Space power and politics | Includes
bibliographical references and index.
Identifiers: LCCN 2018056343 (print) | LCCN 2018060071 (ebook) |
ISBN 9780429755866 (Web PDF) | ISBN 9780429755859 (ePub) |
ISBN 9780429755842 (Mobi) | ISBN 9781138354623 (hardback) |
ISBN 9780429424724 (e-book)
Subjects: LCSH: Space warfare. | Strategy.
Classification: LCC UG1530 (ebook) | LCC UG1530 .K585 2019 (print) |
DDC 358/.84--dc23
LC record available at https://lccn.loc.gov/2018056343

ISBN 13: 978-0-367-67168-6 (pbk)
ISBN 13: 978-1-138-35462-3 (hbk)

Typeset in Times New Roman
by Wearset Ltd, Boldon, Tyne and Wear

For Harvey and Sylvia

Contents

Preface

It was my ambition to write a book that would not be forgotten after two to three years, and that possibly might be picked up more than once by those who are interested in the subject.

Carl von Clausewitz[1]

War and warfare involve both science and art. On the science side, the use of technology, quantitative analysis, measurements, and comparisons—of which Antoine-Henri de Jomini advocated—between friendly and potential adversaries help inform the implementation of strategic, operational, and tactical activities to achieve advantage.[2] On the art side, the human element, policy development, chance, uncertainty, and qualitative assessments likewise affect how relative advantage may be achieved. The title for this work refers to the art aspect of war, commensurate to the context of Sun Tzu's enduring work.[3] Sun Tzu considered the moral, intellectual, and circumstantial elements to war to be more important than the physical aspect and warned not to place reliance on sheer military power.[4] While both science and art have implications when considering the inter-action of dueling states, this work focuses more on advancement of the art of space war and strategy. Science, technology, and quantitative comparisons are relevant in operational and tactical level contingency planning, but it is my view that discussions about the fundamentals of space strategy and the associated art of space warfare are still not well understood.

As illustrated with the quote by Clausewitz at the beginning, it is my intent to present a framework for thinking about war in space that can stand the test of time. By describing space strategy within the context of general theory of war and strategy, it is thought this work will prove useful several years from now. Admittedly, the commercial space sector and associated technology are evolving rapidly. While it is important to frame any theory and strategy of space war in a way that is meaningful and practical given the present environment and state of international affairs, this work will not present a framework that is underpinned by current technology and tactical-level actions. Doing so puts a general theory of space strategy at risk of becoming obsolete and irrelevant soon after being written.

The subject of this work is the theory and strategy of space warfare, which by necessity falls under the general theory of war. Consequently, this work seeks to provide a framework for considering war in the space domain, and does not attempt to "establish a new method of the art of war."[5] As with J.C. Wylie's advice, the intent of this theoretical framework is to provide a point of departure from which to devise, carry out, and criticize strategy.[6] While the lessons of past masters of strategy are presented and referenced, this work seeks to put space war and warfare within the context of the general theory of strategy and provide a cogent foundation for discussing space strategy as a practical matter. There is much about war in space that is presently unknown, because of a scarcity of historical experience of true hostilities and violence in space. Yet there is much we do know about the reasons and intent for waging war in space. Consequently, having a theoretical framework for considering future events in space is considered helpful.

This work was written as a follow-up to the book *Space Warfare: Strategy, Principles and Policy*, which used maritime strategy as inspiration to formulate the strategic principles and theoretical framework for space war and warfare. My previous book served as more of an academic proof, of sorts, to determine if a maritime-inspired space strategy appeared applicable or not. While the principles laid out in *Space Warfare* are still relevant for considering the overarching theory of war in space, this work attempts to provide a broader, more integrated framework for the implementation of space strategy. Moreover, much has happened in the world since the previous book was written. There are less policy and strategy writings using the terms *super powers* and *hegemony*, and the term *great power completion* is now in vogue.

My view of space strategy was shaped by many works, including the works of Sun Tzu (*The Art of War*, c.400–320 BC), Thucydides (*History of the Peloponnesian War*, c.432 BC), Carl von Clausewitz (*On War*, 1815–1827), Antoine Henri de Jomini (*The Art of War*, 1854), Mao Tse-tung (*Selected Military Writing of Mao Tse-tung*, 1926–1957), B.H. Liddell Hart (*Strategy: The Indirect Approach*, 1967), J.C. Wylie (*Military Strategy: A General Theory of Power Control*, 1967), and others that appear in the notes and bibliography.

Furthermore, many of the ideas and concepts discussed were shaped by maritime strategy. I acknowledge the profound contributions of Alfred Thayer Mahan, (including *Naval Strategy Compared and Contrasted with the Principles and Practice of Military Operations on Land*, 1911), Julian S. Corbett (*Some Principles of Maritime Strategy*, 1911), Charles E. Callwell (*Military Operations and Maritime Preponderance*, 1905), Wolfgang Wegener (*The Naval Strategy of the World War*, 1929), Raoul Castex (*Strategic Theories*, 1931–1939), and J.R. Hill's book (*Maritime Strategy for Medium Powers*, 1986).

This work was written for the both the general strategist and space strategist in mind. The implementation of strategy is a practical endeavor, however, so those individuals involved with policy development and contingency planning may find it useful as well. One of the strategist's jobs is to save the people of action both from themselves and the seductive purveyors of the latest all but guaranteed way to win future conflicts.[7] Therefore, this work is meant for those involved in or considering fighting and winning in space.

What is presented here represents my view of space warfare and strategy. It is likely that some will disagree with a one or more of the points made here. Strategy development is rightly so a contestable affair, because the stakes are so high. In the end, the space strategy presented here is what this author considers most relevant and lasting, based upon the revelations of historical experience and the ideas of theorists and strategists that have stood the test of time.

Chapter 1 of this book describes the enduring nature of war—including an overview of the general theory of war and strategy—along with a foundational understanding of the character of space warfare—including the legal regime, normal behaviors, and commercial best practices. Chapter 2 describes the overall framework of a general space strategy, which is shaped by preeminent authors on the general theory of war and strategy. Specific thoughts and principles of the space strategy framework are detailed in this chapter.

In Chapter 3, the impact of ever-advancing technology in warfare is described, along with the implications for the development of space strategy. Historical experience is used to illustrate how new technologies were used initially in warfare with the intent of illuminating those questions strategists should ask about the application of technologies in war in space.

Chapter 4 discusses the interplay between deterrence, dissuasion, and anticipatory self-defense—also known as preemption. The chapter examines the differing perspectives on deterrence and compellence between China, Russia, and the United States. This area is of great concern to this author, because while competing states may not seek direct conflict, they may end up in one due to cultural and strategic misunderstandings.

Building upon the general theory of space war, the next three chapters describe potential space strategies for great, medium, or emerging powers. Chapter 5 describes relevant space strategies for great space powers, which will be relevant when considering competition between China, Russia, and the United States. Chapter 6 details the strategies for medium space powers to gain more international influence and protect national interests, using both non-military and military means. Chapter 7 describes the most effective and efficient strategies for emerging space powers, to include non-governmental organizations, international corporations, insurgents, and terrorists.

The rapidly evolving commercial space sector's role in space strategy is described in Chapter 8, including emerging commercial capabilities and future trends. Finally, Chapter 9 discusses the most pressing considerations moving forward, while acknowledging a strategist's inability to foresee the future. The lessons of historical experience—coupled with the general theory of space war and strategy—are meant to illustrate questions that should be asked by strategists and illuminate what enduring problems will likely remain.

As is frequently the case with a book like this, I am indebted to many who helped and guided me along the way. Included in this long list are: Peter Hays for our daily coffees and for bringing me into the space community fold; James

Vedda for commenting on a complete draft of this work, along with his insights in current space policy issues; Edward Ferguson for our discussions during the writing of our joint articles on space strategy; and Jason Lamb for reading the earliest drafts of several of chapters. For their space strategy writings and ensuing lively debates: Everett Dolman, M.V. Smith, Peter Garretson, and Brent Ziarnick. For those helping shape my understanding of the international community and commercial sector's role in strategy: Dean Cheng, Carissa Christensen, Stuart Eves, Steve Henry, Rebecca Cowen-Hirsch, James Puhek, Jeff Rowlison, Tim Trueheart, and Charity Weeden. To the many unmentioned national security space compatriots who helped shape the ideas here, thank you.

One reason I wrote this book was a result from my teaching at George Washington University's Space Policy Institute. Many thanks to John Logsdon, Scott Pace, Henry Hertzfeld, and John Sheldon for providing me the opportunity to teach space policy courses. For their dauntless research assistance: Nick Boensch, Brennan Bok, and Lexie Weikert. My students constantly remind me how amazing the subject of space is. Moreover, I am especially grateful to Colin Gray, who graciously agreed to be my supervisor while I was pursuing my degree at the University of Reading. His ideas and writings have and continue to inspire many—including this author—to advance the cause of space strategy.

Finally, I would like to thank my family and friends for their patience, as writing this book has consumed what little free time I had during the past year. I am appreciative all the support, assistance, and encouragement in writing this book. That said, I am fully responsible for its contents, conclusions, and any remaining flaws. The views represented here are mine alone and do not necessarily represent those of the U.S. Department of Defense, the George Washington University, or Falcon Research, Inc.

Notes

1 Carl von Clausewitz, "Vorrede," in *Vom Kriege*, erster Band (Berlin: Ferdinand Dümmler, 1832), x.
2 Antoine-Henri de Jomini, *The Art of War* (1862; reprint, London: Greenhill Books, 1992).
3 Sun Tzu, *The Art of War*, trans. Samuel B. Griffith (Oxford: Oxford University Press, 1963).
4 Samuel B. Griffith, preface to *The Art of War*, Sun Tzu, x.
5 Maurice de Saxe, *Reveries on the Art of War*, trans. and ed. Thomas R. Phillips (London: 1757; reprint, Mineola, NY: Dover Publications, 2007), 100.
6 J.C. Wylie, preface to *Military Strategy: A General Theory of Power Control*, with introduction by John B. Hattendorf (New Brunswick, NJ: Rutgers University Press, 1967; reprint, Annapolis, MD: Naval Institute Press, 1989), 2.
7 Colin S. Gray, *Fighting Talk: Forty Maxims on War, Peace, and Strategy* (Westport, CT: Greenwood Publishing, 2007), 42.

Bibliography

Clausewitz, Carl von. *On War*. Translated and edited by Michael Howard and Peter Paret. Princeton, NJ: Princeton University Press, 1989.

Clausewitz, Carl von. *Vom Kriege*, erster Band. Berlin: Ferdinand Dümmler, 1832.

Gray, Colin S. *Fighting Talk: Forty Maxims on War, Peace, and Strategy*. Westport, CT: Greenwood Publishing, 2007.

Griffith, Samuel B. Preface to *The Art of War, ix–xi*. Sun Tzu. Oxford: Oxford University Press, 1963.

Jomini, Antoine-Henri de. *The Art of War*. 1862; reprint, London: Greenhill Books, 1992.

Saxe, Maurice de. *Reveries on the Art of War*. Translated and edited by Thomas R. Phillips. London: 1757; reprint, Mineola, NY: Dover Publications, 2007.

Sun Tzu. *The Art of War*. Translated by Samuel B. Griffith. Oxford: Oxford University Press, 1963.

Wylie, J.C. Preface to *Military Strategy: A General Theory of Power Control, 1–3*. With introduction by John B. Hattendorf. New Brunswick, NJ: Rutgers University Press, 1967; reprint, Annapolis, MD: Naval Institute Press, 1989.

1   Space as a warfighting domain

Polities decide to go to war based on an assessment of fear, honor, and interest, as described by the Athenian historian, strategist, and general Thucydides (c.460 BC–c.400 BC) in his account of the Peloponnesian War.[1] The same will be true of war initiated in or extending into the space. Nowhere in Thucydides' observation does he separate the domains of warfare, which in his day would have been considered the land and sea. Therefore, fear, honor, and interest will play a role prior to initiating conflict in the future, regardless of the domain. Strategists should consider the reasons states or groups decide to use military force, well before the conduct of such actions.

This chapter is divided into two main parts: the enduring nature of war; and the foundational character of space war. The first section will address the nature of war and warfare within the context of the writings of past military theorists and strategists. Next, the foundational underpinnings of activities in space will be discussed, along with the current legal regime, what is considered normal behavior, and today's commercial business practices.

The teachings of the Prussian military strategist and land warfare theorist Carl von Clausewitz (1780–1831) are relevant when considering war in space. Clausewitz's *On War* is considered by many strategists to be the seminal work regarding the general theory of war and warfare. His theory of war offers a framework for reasoning through violent conflict to reach timeless strategic conclusions. He describes a variety of key concepts including the objectives of war being either limited or unlimited; war's limiting factors; and the need to balance violence, chance, and reason. Clausewitz outlines general principles by which to employ offensive and defensive strategies. It is these examinations of the defense and their relative strength compared to the offense that are especially germane in formulating today's space strategy.

In one of his most important observations, Clausewitz underscores that war is an extension of policy by another means. Consequently, war in space should serve political ends as well. As Clausewitz highlights, "The political object—the motive for the war—will thus determine both the military objective to be reached and the amount of effort it requires."[2] The writings of Sun Tzu (544 BC–496 BC) also underscore the role of politics and violence in war. He observed, "War is a matter of vital importance to the State; the province of life and death; the road to survival or ruin."[3] Similarly, Mao Tse-tung—who routinely echoes many of Sun Tzu and Clausewitz's thoughts—comments, "'War is the continuation of politics.' In this sense war is politics and war itself is a political action; since ancient times there has never been a war that did not have a political character."[4]

Because Clausewitz, Sun Tzu, and Mao all comment on the role the state and policy play in war, it is reasonable to expect future military actions in space to have a synergy with policy and politics. War that is initiated in or extended into space must consider the political ends to be achieved and how an effective space strategy can help achieve those ends. Historical experience illustrates that states will frequently protect their interests, no matter where those interests lie. Space will be no exception. For many countries today, political ends and associated national interests span all domains—land, sea, air, space, and cyber. Consequently, the study and conduct of space warfare should not be considered in isolation to the space domain alone.

Because a major state-on-state war has not been initiated in or extended into space, many writers on space warfare imply that such a conflict will be new, without any historical underpinning. It is indeed fortunate that a major state-on-state war in space has not happened. It is speculated that a war extending into space would be devastating to all spacefaring nations. Whether a future conflict in space employs military means that are temporary or permanent, non-kinetic or kinetic, the potential for escalation and resulting orbital debris would likely prove challenging for a return to *antebellum* conditions at the cessation of hostilities. Admittedly, the premise that political ends in space can be linked to military operations is considered controversial to some, as with the topic of weaponization of space in general, but historical experience teaches us to expect linkages between political ends and military operations in all domains in which there are national interests.

The enduring nature of war

> [W]ar is thus an act of force to compel our enemy to do our will.
>
> Carl von Clausewitz[5]

When considering war, and differentiating it from warfare, it has been described that "Warfare is the act of making war. War is a relationship between two states or, in a civil war, two groups. Warfare is only a part of war, although the essential part."[6] So, war is the overall contest between competing states or groups, and warfare is the overall activity in using force and violence to achieve political objectives.

Like Clausewitz, the British maritime strategist Julian Corbett (1854–1922) recognized that both land and sea operations are influenced by national politics and interests. Corbett is renowned for his 1911 work titled *Some Principles of Maritime Strategy* and is acclaimed as Great Britain's greatest maritime strategist.[7] Corbett observed, "War is a form of political intercourse, a continuation of foreign politics which begins when force is introduced to attain our ends."[8] In emphasizing how warfare and national power are intertwined, Corbett declared that the grand strategy of war cannot be decided apart from domestic politics and diplomacy.[9] Accordingly and by logical extension, the military instrument of war is in a vital sense also a political instrument.[10]

War and its conduct are remarkably complex and interdependent activities. In noting the complexity of conflict, Clausewitz notes:

> War is more than a true chameleon that slightly adapts its characteristics to the given case. As a total phenomenon its dominant tendencies always make war a remarkable trinity—composed of primordial violence, hatred and enmity, which are to be regarded as a blind natural force; of the play of chance and probability within which the creative spirit is free to roam; and of its elements of subordination, as an instrument of policy, which makes it subject to reason alone.[11]

While war should be considered an instrument of policy, there is a reciprocal relationship as well.[12] The available military means and what is considered practical will, at times, shape the policy that is sought. Policy may take into consideration what it should aspire to become with an understanding of what it can reasonably achieve militarily.[13] Colin Gray calls this two-way interaction the "strategy bridge."[14]

When considering what is and is not war, it is worth underscoring that it has a deadly component, regardless if the war is great or small, regular or irregular. Clausewitz refers to this deadly aspect as "primordial violence."[15] If hostilities fail to result in the loss of life, it is reasonable to consider other labels or monikers for the event, such as *military action*.[16] While war is not only about violence, that is its distinguishing feature.

Moreover, war and warfare share an enduring, fundamental nature. Jeremy Black explains:

> [I]n its fundamentals war changes far less frequently and significantly than most people appreciate. This is not simply because it involves a constant— the willingness of organized groups to kill and, in particular, to risk death— but also because the material culture of war (the weaponry used and the associated supply systems) which tend to be the focus of attention, is less important than its social, cultural and political contexts and enablers.[17]

Therefore, the nature of war in space will be consistent with a historical understanding of statecraft, strategy, and the use of violence.

Continuum of war and peace

Wars and their conclusions—including the settled peace—matter. Historical experience shows that wars have decided and influenced major decisions, thereby changing the course of world history. While the policy maker and strategist are rightly focused on the conduct of direct and violent military actions, ample thought should be spent on war's conclusion. Frequently, the object in war is to obtain a better peace and not simply to win.[18]

When belligerents consider cessation of hostiles and seek a settled peace, what was gained and lost—including the price paid and sacrifices made in blood and treasure—affects the decision. Edward Luttwak writes:

> Thus war is the origin of peace, by the total victory of one side or another, by sheer exhaustion, or—much more often in history—because the conflict of aims that originally caused the war is resolved by the transformation of aims which war itself brings about: under the impact of its cost in blood, treasure, and agony, the worth of whatever was to be gained, or defended, is reconsidered against its true price, and then ambitions are diminished or renounced.[19]

Consequently, any enduring peace must seem worthwhile to those that have sacrificed so much. Additionally, history has demonstrated that a poorly settled peace can sow the origins for the next war. The defeated have a say in whether peace is eventually reached and if is enduring or temporary. Michael Handel accurately observes:

> For the results of victory to endure, they must be accepted as final by the defeated side, whose interests and concerns must be taken into account. The condition of peace must be such that they appear generous or at least reasonable to the loser. An enduring peace is therefore as reciprocal as everything else in war is.[20]

Without a generous peace from the loser's perspective, the defeated state often considers the outcome merely as a transitory evil that can be remedied through political considerations at some later date.[21]

Friction, chance, and uncertainty

War is—by its very nature—an uncertain affair. According to Clausewitz, war is uncertain because of the friction associated with the complexities of all its activities, including the unpredictability and interplay of violence, chance, and reason, along with the choices made by the enemy. Clausewitz's earliest known use of the term *friction* is to "describe the effect of reality on ideas and intentions in war" and was written in a letter to his future wife, Marie von Brühl.[22] In commenting on the role of chance and uncertainty, Clausewitz writes:

> If we now consider briefly the subjective nature of war—the means by which war has to be fought—it will look more than ever like a gamble....
>
> In short, absolute, so-called mathematical factors never find a firm basis in military calculations. From the very start there is an interplay of possibilities, probabilities, good luck and bad that weaves its way throughout the length and breadth of the tapestry. In the whole range of human activities, war most closely resembles a game of cards.[23]

General friction, as defined by Clausewitz, will continue to be central in future warfare, regardless of technological changes in the means of combat.[24] War has too many drivers and variables to be controlled reliably by either strategists or policy-makers striving to reduce risk and guarantee success. The U.S. naval officer and strategist J.C. Wylie (1911–1993) has similarly observed, "... *we cannot predict with certainty the pattern of war for which we prepare ourselves.*"[25] Alas, there is no such thing as a risk-free or certain path to success.

To deal with friction, chance, and uncertainty, the strategist must be adaptable and flexible over a range of plausible, and some implausible, threats.[26] In noting one of the responsibilities of the strategist, Colin Gray advises, "What the strategist struggles to prevent is the enthronement of the kind of official strategic certainty which precludes the development of strategic and military postural flexibility."[27] Experts and policy makers who advance a knowable and certain future are dangerous to their organizations, as well to those fighting forces who will need to carry out their ill-conceived vision.

Flexibility and adaptability

The execution of a war plan requires flexibility and adaptability. The writings attributed to Sun Tzu emphasize the advantages coming from a practical strategy that is both adaptable and flexible, to exploit any advantages realized during war's conduct. Sun Tzu advises that military leaders should "create situations" that expediently create advantages, to include the use of deception.[28] Perhaps the most illustrative of Sun Tzu's passages calling for flexibility and adaptability is his metaphor of a fighting force being like water, "And as water shapes its flow in accordance with the ground, so an army manages its victory in accordance with the situation of the enemy."[29]

Like Sun Tzu's ideas, B.H. Liddell Hart's concept of the *indirect approach* notes that strategy should adjust as the situation develops in war. Specifically, strategy should not seek solely to overcome the adversary's resistance, but rather should exploit the elements of movement and surprise to achieve victory by throwing the enemy off balance before a potential strike.[30] Seeking to remedy what he considered the calamitous misapplications and misunderstandings of Clausewitzian military thought, Liddell Hart sought a modern vision for strategy relevant to all government activities supporting the attainment of political aims during war, not just the military application of violence.[31] After careful study of past wars, Liddell Hart was persuaded that wars are generally won when the means of war are applied in a way that an opponent is unprepared to meet—or in an indirect fashion.[32] Liddell Hart argues that if a strategist is charged with winning a military victory, then the strategist's:

> responsibility is to seek it under the most advantageous circumstance in order to produce the most profitable result. Hence *his true aim is not so much to seek battle as to seek a strategic situation so advantageous that if it does not of itself produce the decision, its continuation by a battle is sure to*

achieve this. In other words, dislocation is the aim of strategy; its sequel may be either the enemy's dissolution or his easier disruption in battle.[33]

Liddell Hart thought that the situation should be exploited to ensure maximum gains from victory. As a result, the indirect approach includes the application of movement and surprise to dislocate an enemy and exploit any associated military victories to achieve political objectives.[34]

Strategy's purpose

Clausewitz describes the interplay between the leadership, the military, and the populace in determining the will to fight wars.[35] The military strategy of achieving victory is often determined by matching one's own strengths against the enemy's weaknesses, while recognizing that the enemy will attempt to do the same. A good strategy is one that yields sufficient strategic effect to meet the demands of policy and achieve political ends. The strategic effect—the common currency earned by military behavior and action—is generated by the effort, will, and blood of fighting forces in combat.[36]

As used here, *strategy* refers to the art and science of marshalling and directing resources to achieve some objective.[37] Or more simply, it refers to the balancing of one's ends with one's means.[38] When considered within the scope of national interests and policy, it is useful to distinguish strategy further as *grand strategy* or *military strategy*. Grand strategy, also called national strategy, applies during both peace and war to all instruments of national power to achieve a state's objectives. As Liddell Hart explains, "[T]he role of grand strategy—higher strategy—is to co-ordinate and direct all the resources of a nation, or band of nations, towards the attainment of the political object of the war—the goal defined by fundamental policy."[39] In contrast, military strategy typically refers to plans that organize and direct military actions and elements to achieve specific objectives. Below military strategy is battlefield strategy, more commonly referred to as tactics.[40]

Charles Callwell stated that "Strategy is not, however, the final arbiter in war. The battle-field decides," demonstrating his understanding that a practical consideration of the available military instrument dictates what strategies are feasible and what policy choices are achievable.[41] Harold Winton similarly underscores this point, "[W]ar is an intensely practical activity and a ruthless auditor of both individuals and institutions."[42] Strategy's implementation is done at the tactical level of war, so soldiers, marines, sailors, and airmen are the final executers of strategy. Cautioning any would-be strategist, Colin Gray warns, "Officials and soldiers need solutions, not an understanding of complexity bereft of usable answers."[43] Therefore, strategy's development is not an endeavor to be considered in isolation to its practical implementation.[44]

The character of war in space

Although the nature of war is enduring, its character changes. This must be the case, because as Jeremy Black writes, "Every war is unique because of the details of its context (political, social-cultural, economic, technological, military-strategic, geographical, and historical)."[45] Therefore, the nature of war in space will be consistent with a historical understanding of statecraft, strategy, and the use of violence, even though space warfare's character will be different from conflict in other domains. For this reason, space strategy deserves a separate context and lexicon to facilitate a practical understanding and its implementation.[46]

When considering the span of space activities within the United States and many other countries, they are most often divided into four major areas—civil, commercial, intelligence, and military.[47] These four sectors are useful when describing the means available to achieve political and national ends in space.

Civil space activities include government-led efforts to explore space and advance human understanding. The U.S. National Aeronautics and Space Administration and the European Space Agency are two examples of organizations leading such efforts. Civil space includes human and robotic exploration and science missions to advance humanity's understanding of the Earth, solar system, and universe. Of note, civil space efforts may include those of a single country or several countries' civil space programs working together towards a common goal. The U.S. manned Space Shuttle, Soviet Mir, and International Space Station would all fall under civil space activities and programs. Also included in this category are unmanned missions like past Mars robotic rovers and the future James Webb Space Telescope.

Commercial activities are those where companies provide services with the intention of making a profit, whether in the near or long-term. Consequently, commercial space activities are designed to generate wealth from access to and use of space.[48] Today, commercial space capabilities are expanding at an unprecedented rate and are thought to hold the potential to augment a state's national security and provide key elements of its critical national infrastructure. There are several specialized sectors within the commercial space arena, including satellite services, satellite manufacturing, launch industry, and ground equipment. The satellites services sector is perhaps the most diverse and expanding segment, which includes satellite television and radio, satellite broadband, transponder agreements, mobile satellite services (data and voice), and Earth observation services.[49]

The intelligence sector mostly includes intelligence, surveillance, and reconnaissance missions conducted by government agencies for national security purposes. This sector may employ high-resolution imagery considered essential to the formulation and execution of foreign and defense policies.[50] The use of space-based intelligence, surveillance and reconnaissance satellites has helped in the verification of arms control agreements, including those between the United States and the Soviet Union during the height of the Cold War.

Military space activities are those seeking to achieve political objectives through offensive or defensive operations, whether into, through, or from space. Military space activities may include activities and effects to ensure access to and use of space for national purposes. Within current U.S. joint doctrine, there are a variety of military space capabilities and activities to include: space situational awareness; space lift; space control; positioning, navigation, and timing; satellite communications; environmental monitoring; missile warning; and nuclear detonation detection.[51]

Military activities have been contemplated since the dawn of the Space Age and have been pursued by many spacefaring nations. A RAND 1946 report titled "Preliminary Design of an Experimental World-Circling Spaceship" detailed potential military missions in space, which included communications, attack assessment, weather reconnaissance, and strategic reconnaissance missions.[52] The United States conducted a series of high-altitude nuclear detonations designed to confirm that high-energy electrons produced in such detonations would become trapped in the Earth's magnetic field. The Starfish Prime 1.4 megaton detonation, which was 248 miles above Johnson Island in July 1962, caused the premature failure of several satellites in low Earth orbit, as well as significant communications disruption within the vicinity of the Hawaiian Islands.[53] Other past U.S. military activities include anti-satellite testing and the research on the Strategic Defense Initiative program, which was never fielded. The Soviet Union had its own military projects including the Fractional Orbital Bombardment system and anti-satellite testing programs.[54]

To lay a foundational understanding of the expected character of space warfare, the sections that follow describe current views on the legal, just, and safe conduct of space activities in support of national objectives. These topical areas include the current legal regime, the Law of Armed Conflict, rules of engagement, normal behavior in space, and present commercial business practices. From this foundational understanding, the tenets of a general space strategy can be better discerned, as developed in Chapter 2.

Current legal regime

The legal regime for space helps shape the perceptions on how polities' desired political ends compare with the status quo of operations and activities in space. This difference between political ends sought and the status quo is important when deciding to go to war or to seek peace. The great disparities between the sought political ends and the *antebellum* environment shaped by the legal regime may result in a peace being settled reluctantly by the defeated or the peace being only transitory until conflict resumes.

The current legal regime for outer space draws upon both customary international law and international treaty law. Customary international law is often based on hundreds of years of legal precedent and serves as the foundation for observed international law. Customary international law is said to apply to all states. In contrast, international treaty law includes treaties and conventions and binds only those states having ratified a particular treaty or agreement. Although space law spans only about 50 years, it too draws upon customary international law and international treaty law in forming its fundamental precepts.

Of the many treaties and international agreements shaping international behavior in space, the 1967 Outer Space Treaty is still the most relevant and influential. The Outer Space Treaty—more formally known as the Treaty on Principles Governing the Activities of States in the Exploration and Use of Outer Space, Including the Moon and Other Celestial Bodies—came into force on October 10, 1967 and made broad proclamations on how outer space was to be used.[55] Under the Outer Space Treaty, outer space is open to exploration and use by all nations, and space is not subject to national appropriation and should be used for peaceful purposes.

The Outer Space Treaty also addresses legal restrictions on weapons use and some military activities. The treaty prohibits nuclear and other weapons of mass destruction from space.[56] Additionally, military bases, installations, and fortifications may not be erected, nor may weapons tests be undertaken on natural celestial bodies, which include the Earth's Moon but not the Earth. Military personnel, however, may be employed on natural celestial bodies for research and other activities related to "peaceful purposes," including the ability to perform self-defense or denial measures.

Besides the Outer Space Treaty, other treaties and agreements have shaped the legal regime.[57] These other treaties include the 1968 Agreement on the Rescue of Astronauts, the 1972 Liability Convention, the 1975 Registration Convention, the 1972 US/USSR Anti-Ballistic Missile Treaty (of which the United States has since withdrawn), the 1973 International Telecommunication Convention, and the 1980 Convention on the Prohibition of Military or Other Hostile Use of Environmental Modifications Techniques.[58]

The 1973 International Telecommunication Convention is germane when considering space strategy, because the convention has shaped the equitable usage of the electromagnetic frequency spectrum in and through space. The International Telecommunication Union is guided by this convention in overseeing the allocation of operating frequencies used by satellites, because the entire frequency spectrum used by communications satellites is seen as a finite resource. The convention notes,

> In using frequency bands for space radio services Members shall bear in mind that radio frequencies and geostationary satellite orbits are limited natural resources, that they must be used efficiently and economically so that countries or groups of countries may have equitable access to both....[59]

Therefore, the limited natural resources must be allocated through an international organization, thereby ensuring equitable access to and use of space by all countries.[60]

Norms of behavior and "code of conduct"

Other day-to-day activities influence what is considered normal or acceptable in space, even though they are not considered part of the space legal regime. This includes what has been called "norms of behavior" in space. Everett Dolman states, "Norms are standards of behavior defined in terms of rights and obligations."[61] At their most basic, norms are meant to simply describe what is considered normal behavior or activities, given a specific situation or context. It is thought that over time, normative behavior through habituations and expectations of future actions can be made more predictable, affecting the accepted legal regime.[62]

Establishing norms in space is considered by some policy-makers to be a method of enhancing the safety and sustainability of the outer space environment. Additionally, establishing norms of behavior in spaces may appear to be an attractive alternative to modifying and re-negotiating legally binding instruments, including the Outer Space Treaty, because of the ongoing resistance by major space powers to seek changes that could upset the status quo. However, Roger Harrison has noted the downside of trying to establishing norms, in saying, "If specificity is sacrificed to consensus, the resulting regime of non-binding, qualified, and/or vaguely-worded 'norms' may undermine rather than increase stability in space."[63] Among some space policy and security professionals, there is a sense of urgency to address stability and sustainability in outer space through non-legally binding tools and frameworks to "help establish norms for responsible space-faring nations in the near term."[64] It is hoped that norms of behavior in space will enable safety-focused space traffic management, minimize the creation of additional space debris, lead to better conjunction analysis, and help in the sharing of space surveillance information.[65]

Part of the effort to formalize the establishment of norms of behavior in space was the Space Code of Conduct put forth by the European Union. In 2008, the European Union published a draft Code of Conduct for Outer Space Activities. Through voluntary agreement, the Code of Conduct sought member states to establish "policies and procedures to minimize the possibility of accidents ... or any form of harmful interference with other States' right to the peaceful exploration and use of outer space."[66] The code was based on three principles: freedom of access to space for peaceful purposes; preservation of the security and integrity of space objects in orbit; and consideration for the legitimate defensive interests of states. Because the code was not legally binding, there were no formal enforcement mechanisms or substantive recompense to address behavior outside the norms.

The EU Code of Conduct formed the basis for the effort at formalizing an International Code of Conduct, which attempted to garner broader worldwide support. This international effort collapsed in 2015, and some see the failure as a significant diplomatic setback, especially because there is currently no consensus in what manner to reinitiate the effort.[67] Perhaps most disappointing is the fact that most objections to the International Code of Conduct centered on the process for its development rather than on its substance.[68]

While norms and a formal agreement within a code of conduct are useful from a safety and sustainability viewpoint, there is danger in thinking that "norms" will help with identifying a potential adversary's intent. Audrey Schaffer has observed, "Norms are not a panacea for constraining aggressive, hostile, provocative, or otherwise deliberately irresponsible behavior in outer space."[69] Norms will not stop bad actors intent on doing bad things. Deviations from established norms of behavior can help in providing cueing—or indications and warning—when abnormal actions deserve further scrutiny and analysis. Norms should not be used as the threshold for determining hostile intent or a hostile act. Through potential cueing by norm deviation, along with any further scrutiny and resulting analysis, a judgement for action under the inherent right of self-defense per the Standing and Supplemental Rules of Engagement can be better informed.

Commercial business practices and soft norms

Peter Hays and James Vedda have observed that there is a view among many national security space professionals that "Behavioral norms will have a better chance of being accepted and sustained if they're bottom-up rather than top-down."[70] Part of this bottom-up approach for what is considered norms will be how routine, day-to-day space operations are being conducted. The commercial space sector is intimately involved in this process.

The commercial sector's large number of satellites on-orbit—along with associated ground systems—shapes what is considered normal during day-to-day space operations. Carissa Christensen has called these day-to-day operations and associated commercial business practices *soft norms*.[71] These soft norms are mostly not documented, officially agreed upon, or codified in writing.[72] While industry business practices do not necessarily have quantitative metrics or hard thresholds for their use, they are significant nonetheless in shaping the perceptions of what is considered fair, equitable, and safe in space. Because commercial space constitutes a large percentage of on-orbit systems today, commercial space companies are critical in portraying standard day-to-day international and governmental behavior in space.[73] Furthermore, this influence by the commercial space sector is expected to grow in the future because of the projected sector's annual growth rate, which is mostly keeping up with an inflation range of 1–3 percent.[74] This means that the influence of commercial business practices in shaping soft norms is expected to be relevant or even increase. As with other norms, the significance of soft norms is that any deviations from the standard business practices may highlight abnormal behavior or be used in cueing of possible nefarious activity or a potential hostile act.

Moreover, commercial space activities will shape space strategy, because the significant amount of commercial space activities will influence both the political ends and available means for implementing a space strategy. For example, the space-related political ends could include a desire to increase commerce and trade gained through satellites services. The means employed could include using the commercial satellite frequency spectrum for military forces in theater.

Another important consideration is that the current commercial space environment shapes *antebellum* perceptions against which any forced or settled peace will be measured, meaning commercial space helps determine whether any peace is lasting or not.

There are three broad areas for considering commercial space business practices or soft norms. These are minimizing debris and hazards to operations; coordinating rendezvous and proximity operations; and minimizing electromagnetic interference (including coordinating spectrum use). Because the commercial space sector has diverse functions and capabilities, not every commercial space company will be involved in every aspect of these best practices. Nonetheless, these three areas are where most effort should be spent to address safety, sustainability, and minimize risks in space.

Minimizing debris and hazards to operations

To minimize orbital debris and hazards to space operations—whether during launch or on-orbit operations—companies will typically consider designing and manufacturing satellites and launch systems that minimize the amount of generated orbital debris and potential hazards. Prior to commercial launch operations, pre-notification is given to airspace controllers to reduce the hazards to humans in the vicinity launch area, within the airspace of the flight profile, and to deconflict with other launches or orbiting satellites. In low Earth orbit, this includes establishing and implementing a deorbit plan to prevent non-functioning and obsolete satellites from becoming a potential hazard to operational systems. In geostationary orbit, risk mitigation measures may include moving nonfunctioning or obsolete systems to a "graveyard" or other orbit out of the way of functioning satellites. Additionally, endeavors to minimize hazards and risks include prior coordination in gaining a geostationary orbital slot use through International Telecommunications Union. All these efforts are intended to minimize the risks associated with commercial space launch and on-orbit operations. Orbital debris mitigation will be discussed further in Chapter 9.

Coordinating rendezvous and proximity operations

When commercial satellite operators are considering rendezvous or proximity operations with other on-orbit satellites, they will typically want to coordinate any maneuver planning with other potentially-affected operators, as deemed necessary. This coordination may include sharing conjunction analysis and associated risks of close proximity. Upon occasion, when a satellite malfunctions or becomes inoperable, commercial companies will communicate to commercial or governmental organizations the situation, conjunction analysis, and any associated risks.

To facilitate the sharing of commercially available information, companies like AGI have established a Commercial Space Operations Center to use a global network of commercial sensors to track satellites and generate a comprehensive space situational awareness picture.[75] Closely related to efforts to minimize hazards on-orbit and share conjunction analysis through space situational awareness is Space Traffic Management activities. Space Traffic Management (STM) objectives are said to include enhancing, facilitating, and supporting continued development of the commercial space industry, ensuring safe commercial space operations, minimizing false alarms of conjunction hazards, and fostering development and sharing of norms of behavior and best practices.[76] Within the United States, the current plan is for the Department of Commerce to lead governmental STM activities.[77] More on STM will be examined in Chapter 9.

Minimizing electromagnetic interference

It is in the interest of commercial space operators and service providers to minimize electromagnetic inference and coordinate frequency use across the spectrum. Commercial service providers will commonly gain approval for on-orbit radio spectrum use through a domestic regulatory body such as the U.S. Federal Communications Commission and then seek approval through the International Telecommunications Union. The Space Data Association—a non-profit association of satellite operators—seeks controlled, reliable, and efficient data sharing to improve the safety and integrity of the space environment and the radio frequency spectrum.[78] The intent of these efforts includes coordinating spectrum use with other owners/operators. This coordination of spectrum involves minimizing risks from satellite interference due to the reckless use of lasers, such as those used for range finding.

Looking forward, the commercial space industry will continue to shape what is considered normal operations in space. This includes the insurance industry reinforcing any de facto norms through the insurance underwriting process and pricing for launch providers or on-orbit operators. Typically, the insurance industry sets prices based on what is considered safe in comparison to best practices of the entire commercial space enterprise. Consequently, the underwriting process serves as one of the key instruments shaping launch and on-orbit operations.

Last, there are other future challenges that the commercial sector will help address or shape through establishing soft norms. The future challenges include: how commercial companies can use orbital debris—such as rocket bodies—for economic gain; proximity operations, such as when permission is required to image or approach another's satellite; on-orbit servicing, to include refueling and inspection; how to promote human safety in future space tourism activities; and how space traffic management will be coordinated and integrated internationally. The commercial space sector will be examined in more detail in Chapter 8.

Conclusion

This first chapter lays a foundational perspective on today's considerations for the access to and use of space, along with how conflict in space fits within the enduring nature of war and historical experience. Polities will decide to initiate a war in, or extending into, space based upon an assessment of fear, honor, and interest. The character of war in space, however, will be different than conflict in the other domains, and the current legal regime, normal behavior in space, and commercial best practices help shape this character. The present state of governmental and commercial activities in space has an impact on the formulation of a practical space strategy, as these activities will directly influence the political ends sought and military means employed.

What follows in the next chapter is a theoretical framework of space strategy. This theory is subordinate to the general theory of war, as laid out by Clausewitz and other masters of military theory and strategy.

Notes

1   Robert B. Strassler, *The Landmark Thucydides: A Comprehensive Guide to the Peloponnesian War* (New York: Free Press, 1996), 43.
2   Carl von Clausewitz, *On War*, trans. and eds. Michael Howard and Peter Paret (Princeton, NJ: Princeton University Press, 1989), 81.
3   Sun Tzu, *The Art of War*, trans. Samuel B. Griffith (Oxford: Oxford University Press, 1963), 63.
4   Mao Tse-tung, "On Protracted War," in *Selected Military Writings of Mao Tse-tung* (Peking: Foreign Language Press, 1963), 226–227.
5   Carl von Clausewitz, *Vom Kriege*, erster Band (Berlin: Ferdinand Dümmler, 1832), 4.
6   Peter Browning, *The Changing Nature of Warfare: The Development of Land Warfare from 1792 to 1945* (Cambridge: Cambridge University Press, 2002), 2.
7   Eric J. Grove, introduction to *Some Principles of Maritime Strategy*, Julian S. Corbett (London: Longmans, Green and Co., 1911; reprint, Annapolis, MD: Naval Institute Press, 1988), xxxvi. Comments attributed to the London Times.
8   Julian S. Corbett, *Some Principles of Maritime Strategy* (London: Longmans, Green and Co., 1911; reprint, Annapolis, MD: Naval Institute Press, 1988), 307.
9   Corbett, *Some Principles of Maritime Strategy*, 308.
10 Colin S. Gray, *Airpower for Strategic Effect* (Maxwell Air Force Base, AL: Air University Press, 2012), 7.
11 Clausewitz, *On War*, 189.
12 Colin S. Gray, *Fighting Talk: Forty Maxims on War, Peace, and Strategy* (Westport, CT: Greenwood Publishing, 2007), 29.
13 Ibid.
14 Colin S. Gray, *The Strategy Bridge: Theory for Practice* (Oxford: Oxford University Press, 2010).
15 Clausewitz, *On War*, 189.
16 This does not discount that war's conduct may include coercion through intimidation or threat of military force.
17 Jeremy Black, *War and the New Disorder in the 21st Century* (New York: Continuum, 2004), 163–164.
18 B.H. Liddell Hart, *Strategy: The Indirect Approach*, 2nd ed. (London: Faber and Faber, 1967), 366.
19 Edward N. Luttwak, *Strategy: The Logic of War and Peace* (Cambridge, MA: Harvard University Press, 1987), 58.
20 Michael I. Handel, *Masters of War: Classical Strategic Thought*, 3rd ed. (London: Frank Cass, 2001), 197–198.
21 Clausewitz, *On War*, 80.

22 Quoted in Barry D. Watts, "Clausewitzian Friction and Future War," McNair Paper No. 68 (National Defense University, 2004), 1.
23 Clausewitz, *On War*, 85–86.
24 Watts, "Clausewitzian Friction and Future War," 83.
25 J.C. Wylie, *Military Strategy: A General Theory of Power Control*, with introduction by John B. Hattendorf (New Brunswick, NJ: Rutgers University Press, 1967; reprint, Annapolis, MD: Naval Institute Press, 1989), 70. Emphasis original.
26 Gray, *Fighting Talk*, 160.
27 Ibid.
28 Sun Tzu, *The Art of War*, 66.
29 Ibid., 101.
30 Liddell Hart, *Strategy: The Indirect Approach*, 337.
31 Ibid., 366.
32 Ibid., 25.
33 Ibid., 339; emphasis original.
34 Liddell Hart, *Strategy: The Indirect Approach*, 349.
35 "The first … mainly concerns the people; the second the commander and his army; the third the government." Clausewitz, *On War*, 89.
36 Gray, *Fighting Talk*, 55.
37 John J. Klein, *Space Warfare: Strategy, Principles and Policy* (Abingdon: Routledge, 2006), 4.
38 Strategy and Force Planning Faculty, *Strategy and Force Planning*, 3rd ed. (Newport, RI: Naval War College Press, 2000), 20. For another description of strategy, see John M. Collins, *Grand Strategy: Principles and Practices* (Annapolis, MD: Naval Institute Press, 1973), 14.
39 Liddell Hart, *Strategy: The Indirect Approach*, 335–336.
40 Other divisions of warfare exist. "The theory of military art, as applied to military operations of various scope, is divided into strategy, operational art, and tactics." V.D. Sokolovskii, *Soviet Military Strategy*, trans. Herbert Dinerstein, Leon Gouré and Thomas Wolfe (Englewood Cliffs, NJ: Prentice-Hall, 1963), 88.
41 Charles E. Callwell, *Small Wars: A Tactical Textbook for Imperial Soldiers* (London, 1906; reprint, London: Greenhill Book, 1990), 90.
42 Harold R. Winton, "On the Nature of Military Theory," in *Toward a Theory of Spacepower: Selected Essays*, eds. Charles D. Lutes and Peter L. Hays, with Vincent A. Mazo, Lisa M. Yambrick, and M. Elaine Bunn (Washington: National Defense University Press, 2011), 23.
43 Gray, *Fighting Talk*, 5.
44 Bernard Brodie, *War and Politics* (New York: The Macmillan Company, 1973), 452.
45 Jeremy Black, *Rethinking Military History* (Abingdon: Routledge, 2004), 19.
46 Klein, *Space Warfare*, 154.
47 Peter L. Hays, James M. Smith, Alan R. Van Tassel, and Guy M. Walsh, *Spacepower for a New Millennium: Space and National Security* (New York: McGraw-Hill, 2000), 2–3. Some have also included the "international sector" on the list. See Dana Johnson, Scott Pace, and C. Bryan Gabbard, *Space: Emerging Options for National Power* (Santa Monica, CA: RAND Corporation, 1988), 18.
48 Hays et al., *Spacepower for a New Millennium*, 2.
49 "2016 State of the Satellite Industry Report" (Bryce Space and Technology and Satellite Industry Association, September 2016), https://brycetech.com/downloads/SIA_SSIR_2016.pdf
50 Hays et al., *Spacepower for a New Millennium*, 2.
51 Joint Chiefs of Staff, *Space Operations*, Joint Publication 3–14 (April 10, 2018), ix, www.jcs.mil/Portals/36/Documents/Doctrine/pubs/jp3_14.pdf.
52 Hays et al., *Spacepower for a New Millennium*, 2.
53 Ibid., 17.

54 Ibid., 30 and 35. *See also* Paul B. Stares, *The Militarization of Space: U.S. Policy, 1945–1984* (Ithaca, NY: Cornell University Press, 1985), 80.

55 United Nations General Assembly, resolution 2222 (XXI), *Treaty on Principles Governing the Activities of States in the Exploration and Use of Outer Space, including the Moon and Other Celestial Bodies*, or *The Outer Space Treaty* (1967), www.unoosa.org/oosa/en/ourwork/spacelaw/treaties/outerspacetreaty.html

56 Ibid., Article IV.

57 Nathan C. Goldman, *American Space Law: International and Domestic*, 2nd ed. (Ames, IA: Iowa State University Press, 1996), 68. The Outer Space Treaty is considered the most definitive on the subject, but it owes much of its content to the 1963 Declaration of Legal Principles Governing the Activities of States in the Exploration and Use of Outer Space and the 1959 Antarctic Treaty.

58 Ibid. See appendices 1, 2, 3, and 4. Although the administration of George W. Bush formally announced in December 2001 the withdrawal of the United States from the Anti-Ballistic Missile Treaty, and officially did so in June 2002, the treaty still served to shape current perceptions on the uses of space.

59 U.S. Congress, House Committee on Commerce, Science, and Transportation, *Space Law: Selected Basic Documents*, 2nd ed. (95th Congress, 2nd session, 1978), 86–87.

60 Goldman, *American Space Law*, 28–29.

61 Everett C. Dolman, *Astropolitik: Classical Geopolitics in the Space Age* (London: Frank Cass, 2002), 87.

62 Ibid., 87.

63 Roger G. Harrison, Key Points, in "Space and Verification. Vol. I: Policy Implications" (Eisenhower Center for Space and Defense Studies, 2010), https://swfound.org/media/37101/space%20and%20verification%20vol%201%20-%20policy%20implications.pdf

64 Michael Krepon, "Origins of and rationale for a space code of conduct," in *Decoding the International Code of Conduct for Outer Space Activities*, ed. Ajey Lele (New Delhi: Pentagon Security International, 2012), 31.

65 European Union, "International Code of Conduct for Outer Space Activities," draft version 31 (2014), 7, https://eeas.europa.eu/sites/eeas/files/space_code_conduct_draft_vers_31-march-2014_en.pdf

66 Ibid.

67 James A. Vedda and Peter L. Hays, "Major Policy Issues in Evolving Global Space Operations" (The Mitchell Institute of Aerospace Studies, February 2018), 21, www.aerospace.org/publications/policy-papers/major-policy-issues-in-evolving-global-space-operations/

68 Ibid.

69 Audrey Schaffer, "The Role of Space Norms in Protection and Defense," *Joint Force Quarterly* 87 (October 2017), 89, http://ndupress.ndu.edu/Publications/Article/1325996/the-role-of-space-norms-in-protection-and-defense/

70 Vedda and Hays, "Major Policy Issues in Evolving Global Space Operations," 2.

71 Carissa Christensen, "Commercial Space and Soft Norms" (presentation, Bryce Space and Technology, London, 2017).

72 Ibid.

73 "2017 State of the Satellite Industry Report" (Bryce Space and Technology and Satellite Industry Association, June 2017), 9, www.sia.org/wp-content/uploads/2017/07/SIA-SSIR-2017.pdf

74 Ibid.

75 See "ComSpOC," AGI, accessed August 1, 2018, www.agi.com/comspoc

76 Vedda and Hays, "Major Policy Issues in Evolving Global Space Operations," 3.

77 "President Signs Space Traffic Management Policy," Department of Commerce, Office of Space Commerce, press release, June 18, 2018, www.space.commerce.gov/president-signs-space-traffic-management-policy/

78 "Space Data Association," Space Data Association, accessed August 1, 2018, www.space-data.org

Bibliography

"2016 State of the Satellite Industry Report." Bryce Space and Technology and Satellite Industry Association, September 2016. https://brycetech.com/downloads/SIA_SSIR_2016.pdf

"2017 State of the Satellite Industry Report." Bryce Space and Technology and Satellite Industry Association, June 2017. www.sia.org/wp-content/uploads/2017/07/SIA-SSIR-2017.pdf

Black, Jeremy. *Rethinking Military History*. Abingdon: Routledge, 2004.

Black, Jeremy. *War and the New Disorder in the 21st Century*. New York: Continuum, 2004.

Brodie, Bernard. *War and Politics*. New York: The Macmillan Company, 1973.

Browning, Peter. *The Changing Nature of Warfare: The Development of Land Warfare from 1792 to 1945*. Cambridge: Cambridge University Press, 2002.

Callwell, Charles E. *Small Wars: A Tactical Textbook for Imperial Soldiers*. London, 1906; reprint, London: Greenhill Book, 1990.

Christensen, Carissa. "Commercial Space and Soft Norms." Presentation. Bryce Space and Technology, London, 2017.

Clausewitz, Carl von. *On War*. Translated and edited by Michael Howard and Peter Paret. Princeton, NJ: Princeton University Press, 1989.

Clausewitz, Carl von. *Vom Kriege*, erster Band. Berlin: Ferdinand Dümmler, 1832.

Collins, John M. *Grand Strategy: Principles and Practices*. Annapolis, MD: Naval Institute Press, 1973.

"ComSpOC." AGI. Accessed August 1, 2018, www.agi.com/comspoc

Corbett, Julian S. *Some Principles of Maritime Strategy*. London: Longmans, Green and Co., 1911; reprint, Annapolis, MD: Naval Institute Press, 1988.

Dolman, Everett C. *Astropolitik: Classical Geopolitics in the Space Age*. London: Frank Cass, 2002.

European Union. "International Code of Conduct for Outer Space Activities." Draft version 31. 2014. https://eeas.europa.eu/sites/eeas/files/space_code_conduct_draft_vers_31-march-2014_en.pdf

Goldman, Nathan C. *American Space Law: International and Domestic*. 2nd edition. Ames, IA: Iowa State University Press, 1996.

Gray, Colin S. *Airpower for Strategic Effect*. Maxwell Air Force Base, AL: Air University Press, 2012.

Gray, Colin S. *The Strategy Bridge: Theory for Practice*. Oxford: Oxford University Press, 2010.

Gray, Colin S. *Fighting Talk: Forty Maxims on War, Peace, and Strategy*. Westport, CT: Greenwood Publishing, 2007.

Grove, Eric J. Introduction to *Some Principles of Maritime Strategy*, xi–xlv. Julian S. Corbett. London: Longmans, Green and Co., 1911; reprint, Annapolis, MD: Naval Institute Press, 1988.

Handel, Michael I. *Masters of War: Classical Strategic Thought*. 3rd edition. London: Frank Cass, 2001.

Harrison, Roger G. "Space and Verification. Vol. I: Policy Implications." Eisenhower Center for Space and Defense Studies, 2010. https://swfound.org/media/37101/space%20and%20verification%20vol%201%20-%20policy%20implications.pdf

Hays, Peter L., James M. Smith, Alan R. Van Tassel, and Guy M. Walsh. *Spacepower for a New Millennium: Space and National Security*. New York: McGraw-Hill, 2000.

Johnson, Dana, Scott Pace, and C. Bryan Gabbard. *Space: Emerging Options for National Power*. Santa Monica, CA: RAND Corporation, 1988.

Joint Chiefs of Staff. *Space Operations*. Joint Publication 3–14. April 10, 2018. www.jcs. mil/Portals/36/Documents/Doctrine/pubs/jp3_14.pdf.

Klein, John J. *Space Warfare: Strategy, Principles and Policy*. Abingdon: Routledge, 2006.

Krepon, Michael. "Origins of and Rationale for a Space Code of Conduct." In *Decoding the International Code of Conduct for Outer Space Activities*, edited by Ajey Lele, 30–34. New Delhi: Pentagon Security International, 2012.

Liddell Hart, B.H. *Strategy: The Indirect Approach*. 2nd edition. London: Faber and Faber, 1967.

Luttwak, Edward N. *Strategy: The Logic of War and Peace*. Cambridge, MA: Harvard University Press, 1987.

Mao Tse-tung. *Selected Military Writings of Mao Tse-tung*. Peking: Foreign Language Press, 1963.

"President Signs Space Traffic Management Policy." Department of Commerce. Office of Space Commerce. Press release, June 18, 2018. www.space.commerce.gov/president-signs-space-traffic-management-policy/

Schaffer, Audrey. "The Role of Space Norms in Protection and Defense." *Joint Force Quarterly* 87 (October 2017): 88–92. http://ndupress.ndu.edu/Publications/Article/1325996/the-role-of-space-norms-in-protection-and-defense/

Sokolovskii, V.D. *Soviet Military Strategy*. Translated by Herbert Dinerstein, Leon Gouré and Thomas Wolfe. Englewood Cliffs, NJ: Prentice-Hall, 1963.

"Space Data Association." Space Data Association. Accessed August 1, 2018, www. space-data.org

Stares, Paul B. *The Militarization of Space: U.S. Policy, 1945–1984*. Ithaca, NY: Cornell University Press, 1985.

Strassler, Robert B. *The Landmark Thucydides: A Comprehensive Guide to the Peloponnesian War*. New York: Free Press, 1996.

Strategy and Force Planning Faculty. *Strategy and Force Planning*. 3rd edition. Newport, RI: Naval War College Press, 2000.

Sun Tzu. *The Art of War*. Translated by Samuel B. Griffith. Oxford: Oxford University Press, 1963.

United Nations General Assembly. Resolution 2222 (XXI). *Treaty on Principles Governing the Activities of States in the Exploration and Use of Outer Space, including the Moon and Other Celestial Bodies*, or *The Outer Space Treaty*. 1967. www.unoosa.org/oosa/en/ourwork/spacelaw/treaties/outerspacetreaty.html

U.S. Congress, House Committee on Commerce, Science, and Transportation. *Space Law: Selected Basic Documents*. 2nd edition. 95th Congress, 2nd session, 1978.

Vedda, James A. and Peter L. Hays. "Major Policy Issues in Evolving Global Space Operations." The Mitchell Institute of Aerospace Studies, February 2018. www.aero space.org/publications/policy-papers/major-policy-issues-in-evolving-global-space-operations/

Watts, Barry D. "Clausewitzian Friction and Future War." McNair Paper No. 68. National Defense University, 2004.

Winton, Harold R. "On the Nature of Military Theory." In *Toward a Theory of Spacepower: Selected Essays*, edited by Charles D. Lutes and Peter L. Hays, with Vincent A.

Mazo, Lisa M. Yambrick, and M. Elaine Bunn. Washington: National Defense University Press, 2011.

Wylie, J.C. *Military Strategy: A General Theory of Power Control*. With introduction by John B. Hattendorf. New Brunswick, NJ: Rutgers University Press, 1967; reprint, Annapolis, MD: Naval Institute Press, 1989.

2 Space strategy

Everything is very simple in war, but the simplest thing is difficult.

Clausewitz[1]

Any general theory for space strategy must complement the general theory of warfare—which is domain agnostic. Although needing to fall under the overall general theory for war, the formulation of an enduring space strategy necessitates that its character be distinct from land, maritime, air, and cyber strategies. Yet actions and desired effects for space strategy may cross through these other domains; consequently, space strategy will, at times, augment the strategies in the other domains.

The strategic framework for space strategy that follows is meant to complement the general theory of war as described by past strategists' work that have stood the test of time, as described in Chapter 1. This chapter will describe the purpose of space strategy, along with various tenets of a general theory of space strategy—including command of space, offensive and defensive strategies, strategy for the less capable, dispersal and concentration, and positions. Many of these tenets and concepts will apply to the strategies of great, medium, and emerging space powers, which are discussed in Chapters 5, 6, and 7 respectively.

The principles that follow are gleaned using strategic analogies to land, sea, or air warfare. John Sheldon and Colin Gray rightly note the limitations of using strategic analogies in developing a general framework for space strategy. They observe that the use of strategic analogies is a "necessary step on the road to creating and developing an enduring and universal theory of spacepower."[2] Eventually, however, the crutch of strategic analogies will need to be kicked away to make better progress in theory-making through inductive reasoning.[3] Despite the failings of using strategic analogies, the framework provided here—being gleaned through the enduring nature of war and lessons of historical experience—is meant to help "educate the mind of the future commander, or, more accurately, to guide him in his self-education...."[4]

The purpose of space strategy

The purpose of space strategy is to ensure access to and use of space.[5] This stated purpose of space strategy is gleaned from many documented space strategies, along with analogous comparisons to air and maritime strategies. While some strategists may take exception with this statement or view it as debatable—taken as a whole—this statement on space strategy is thought by this author to be useful, practical, and correct.

Some strategists have extrapolated space strategy from the theories of air and maritime warfare.[6] Early air power thinking recognized the need to access and use the air domain, along with the movement of things throughout it. Just prior to the onset of World War I, Giulio Douhet (1869–1930) concluded, "To have command of the air means to be in a position to prevent the enemy from flying while retaining the ability to fly oneself."[7] In an early definition of airpower, U.S. Army general and early air power theorist William "Billy" Mitchell (1879–1936) said,

> Air power may be defined as the ability to do something in the air. It consists of transporting all sorts of things by aircraft from one place to another, and as air covers the whole world there is no place that is immune from influence by aircraft.[8]

While the U.S. naval officer and strategist Alfred Thayer Mahan (1840–1914) promoted the virtues of offensive actions and seeking the decisive battle, he discussed the importance of maritime communications—or those lines of communication typically used for supply and trade, but also those of a strategic nature and critical for a state's survival. Mahan wrote, "Communications dominate war; broadly considered, they are the most important single element in strategy, political or military."[9] Without access to such lines of communication, offensive actions at sea are destined to fail.

More recent documents reflect a similar underlining philosophy about the primacy of ensuring access and use. According to Chinese security analysts, the goal of China's space operations and strategy is to achieve *space superiority*, defined as "ensuring one's ability to fully use space while at the same time limiting, weakening, and destroying an adversary's space forces."[10] Similar in its conclusion on the importance of space, the 2017 U.S. National Security Strategy notes, "The United States considers unfettered access to and freedom to operate in space to be a vital interest."[11] The European Union's 2016 Space Strategy for Europe echoes the idea as well: "Europe needs to ensure its freedom of action and autonomy. It needs to have access to space and be able to use it safely."[12] Taken as a whole, analogies to air and maritime strategies—along with space policy and strategies of major space powers—exemplify that for the general theory of space strategy, the purpose is to ensure access to and use of space.

Perhaps the best example highlighting the importance of accessing and using space is a 2015 U.S. white paper titled *Space Domain Mission Assurance: A*

Resilience Taxonomy.[13] This paper details thinking within the U.S. national security space enterprise on mission assurance and resiliency. The white paper defines *mission assurance* as:

> A process to protect or ensure the continued function and resilience of capabilities and assets—including personnel, equipment, facilities, networks, information and information systems, infrastructure, and supply chains—critical to the performance of DoD mission essential functions in any operating environment or condition.[14]

In defining *resilience*, the document says it includes six discrete characteristics: disaggregation, distribution, diversification, protection, proliferation, and deception.[15] Also, resilience is said to include the "ability of an architecture to support the functions necessary for mission success with higher probability, shorter periods of reduced capability, and across a wider range of scenarios, conditions, and threats, in spite of hostile action or adverse conditions....."[16] Although the white paper attempts to be inclusive of many disparate ideas, the definitions of mission assurance and resilience imply the underlying need to ensure access to and use of space.

Achieving space strategy's purpose is enabled through celestial lines of communication (CLOCs).[17] As defined here, CLOCs are those lines of communication in, through, and from space used for the movement of trade, materiel, supplies, personnel, spacecraft, electromagnetic transmissions, and some military effects.[18] By ensuring access to its own CLOCs, a state can help protect diplomatic, economic, informational, and military interests. Because a spacefaring nation's access to and use of space is vitally important, space strategy must consider the protection and defense of one's own CLOCs, while limiting the adversary's ability to access or use its CLOCs.

As with sea lines of communication, lines of communication in space often run parallel to, and are frequently shared with, an adversary. This sharing is exemplified by potential competitors having access to the same radio frequency spectrum and orbital regimes. Consequently, an adversary's space communications frequently cannot be attacked without affecting those of both parties, and even neutral parties. For this reason, ample consideration should be given prior to the initiation of military action against an adversary as to whether the result will be a significant denial or obstruction of one's own access to and use of space, or negatively impacting neutral parties and commercial space partners. As has been often written about, on-orbit debris generation or electromagnetic inference is a concern not only among competing space powers, but for the entire the international space community.

Consistent with historical experience, states may protect their national security interests, no matter where they lie. Space is no exception. Although power and influence are often discussed in the context of internationally recognized states, non-state actors—to include non-governmental organizations and commercial companies—can achieve some strategic effects through their space activities as well. Consequently, space strategy and its practical implementation is relevant to states and organizations, both great and small.

Command of space

The inherent value of space is what it allows you to do. Because of this, those with interests in space may attempt to preserve and promote their continued access to and use of space. This preservation and promotion are accomplished through the concept of command of space. Command of space entails the ability to ensure access to and use of CLOCs when needed to support the instruments of national power.[19] It also includes the capability to prevent or deny the enemy's access to and use of his CLOCs, or at least minimize the most severe consequences an adversary can deliver along them.

Bleddyn Bowen has concluded that space warfare only has meaning so far as it works towards the command of space and that the command of space is about manipulating celestial lines of communication.[20] This idea of establishing a level of command of space is commensurate with classical maritime strategy. Julian Corbett, in discussing the purpose of maritime strategy, wrote, "The object of naval warfare must always be directly or indirectly either to secure the command of the sea or to prevent the enemy from securing it."[21]

U.S. defense department doctrine on space incorporates the fundamental premise of command of space, albeit using the terminology of *control* instead. The 2018 U.S. joint doctrine titled *Space Operations* states, "US space forces conduct space control to ensure freedom of action in space for the US and its allies, and when directed, to deny an adversary freedom of action in space."[22] So, like command of space, U.S. joint doctrine notes the need to protect freedom of action for U.S. space assets, while denying an adversary's use of its space assets.

Additional parallels between command of space and control are found in J.C. Wylie's writings. J.C. Wylie (1911–1993), a U.S. naval officer and accomplished strategist, described a general theory of strategy, to include his idea of *control*. In his book *Military Strategy*, he wrote:

> The primary aim of the strategist in the conduct of war is some selected degree of control of the enemy for the strategist's own purpose; this is achieved by control of the pattern of war; and this control of the pattern of war is had by manipulation of the center of gravity of war to the advantage of the strategist and the disadvantage of the opponent.
>
> The successful strategist is the one who controls the nature and the placement and the timing and the weight of the centers of gravity of war, and who exploits the resulting control of the pattern of war toward his own ends.[23]

Therefore, Wylie recognizes that the strategist's role is not establishing control in an absolute sense, but to some degree.

While there are no reasons why space operations cannot deliver decisive strategic effects to achieve success, the conditions allowing for such a victory should be considered rare, indeed. This is because to have the greatest impact and affect the strategic-level of war, conflict must affect the preponderance of people where they live. Colin Gray says it best, "All conflict must have terrestrial reference because man can live only upon the land...."[24] As a result, there will be practical limits to what space operations can achieve strategically, no matter how significant a level of command of space is achieved or how well operations are executed. Command of space may achieve strategic effect, but tactical and operational space actions will only in the rarest occasions be inherently strategic. This is because strategic effect is decided by the target and not by the means of attack.[25]

General and local

Because space is vast, where and when command of space is gained and exercised is important. Consequently, command can be differentiated based on where and for how long it is achieved. For space warfare, this means command can be either general or local, and either persistent or temporary.[26]

General command of space is achieved when an enemy is no longer able to act in a significant or dangerous way against a state's use of CLOCs, and the enemy cannot adequately defend his own CLOCs. With only minor exceptions, general command allows the unfettered use of space for diplomacy, trade, commerce, informational services, or military operations. Drawing upon the similarities to maritime strategy, general command is achieved when the enemy is no longer able to "act dangerously" against the lines of passage and communications or even to defend his own; as a result, the enemy is unable to interfere seriously against trade, military, or diplomatic activities.[27]

For local command, command is gained or exercised within a region that is less than the total region wherein a state's interests lie. Local command is less than optimal, but it is a suitable recourse for less capable space powers. Gaining local command allows for the protection of a state's most important CLOCs through offensive or defensive measures, while its adversary potentially cannot use its communications within that same limited region. The reasons for striving to attain local command can include the desire to gain prestige among the international community, garner domestic political support, protect economic interests, or gain a relative military advantage within a specific region of space.

Persistent and temporary

Not only can command be general or local, it may also be either persistent or temporary. Persistent command means that, despite the adversary's attempts, the element of time is no longer a significant strategic factor in the execution of warfare into, from, and through space. When command is both general and persistent, it does not mean the enemy will not act, but that it is severely weakened to a point where his efforts are unlikely to affect the war's outcome at the strategic level. When command is local and persistent, it means that the CLOCs are protected within a specified region for the near future, but the outcome of the war or conflict is still not ensured.

Temporary command means that either general or local command is gained for a specific period to achieve either military or non-military objectives. A less capable space force can often achieve command that is local and temporary by concentrating assets where its opponent is not, and it can also be achieved by taking a sizable defensive posture during a period or within a certain region. Such a defensive posture can prevent a stronger space power from operating uncontested within a specific region.

Current Chinese thinking on space strategy is in line with this view of command being based on time and location. China's military writings describe counter space capabilities being used to conduct actions that the People's Liberation Army (PLA) calls "space attack and defense operations." According to a PLA Academy of Military Science strategy document, the objective of Chinese space attack and defense operations is to achieve space superiority within a certain period and at a certain location.[28]

A barrier to action

From an understanding of command of space and the different types of command that may be achieved, it is understood that CLOCs are readily accessible to those who exercise command; however, space becomes a "barrier" to those who cannot.[29] Because outer space is an unnatural environment for people to live in, moving along or using CLOCs is not a simple matter. The ability to access and use these lines of communication is paramount, and only by doing so can the advantages of operating in space be realized. Those with the strongest and most effective command of space are more easily able to move materiel, trade, supplies, personnel, spacecraft, military effects, data, and information along CLOCs; however, to those without an adequate level of command, these same lines of communication are more likely to become an obstacle to such access and use. Because space is interdependent with other warfare domains, space can be made a barrier by any combination of land, sea, air, cyber, or space actions.

There are generally three intentions for wanting to make space a barrier.[30] First, it can be for defensive intent, such as wanting to deny the enemy's capability to launch an overwhelming surprise attack. Second, it can be when there is limited intent. This would be the case when one state plans on conducting hostile actions for limited aims against another, but also desires to prevent the enemy's capability to escalate the conflict. Third, the intent can be when a state is conducting an unlimited war with unlimited objectives, such as the unconditional surrender of the enemy, and desires to prevent the enemy's unlimited counter-attack into, through, or from space. Best summarizing the value of making space as a barrier to action is a paraphrase of Francis Bacon (1561–1626), a prominent Elizabethan and Jacobean politician, philosopher, and writer: "He that commands space is at great liberty and may take as much or as little space warfare as he will."[31]

Command of space—including making it a barrier—may be achieved without the overt use of military force, such as by applying the non-military instruments of national power. Whether using diplomatic, economic, or informational measures— or what has been called by Joseph Nye as *soft power*—pressure and influence can be applied to coerce a foe to choose not to develop terrestrial and space-based weapons.[32] Such a conscious choice by a foe may be effectively equivalent to the end state reached through traditional military methods. Although the end state is the same—the enemy not using space as a medium of attack—the method of achieving it is substantially different. Some of the greatest advantages are realized through these non-military means because diplomatic, economic, and informational measures are frequently reversible in nature and may not cause irreparable harm to another or to the space environment. This approach to achieving political ends is in line with Sun Tzu's belief that it is better to win without fighting.[33]

Angels and demons

Command of space and implementing a practical space strategy will necessitate specific space systems that are designed and employed for assuring one's own access to and use of space. Moreover, some systems will be needed to impact negatively an adversary's ability to access and use space. Such systems will need to operate along CLOCs and where space communications tend to congregate, but they will also need to disperse along the most extensive lines of communication to support mission assurance and resiliency efforts. These systems—whether terrestrial or space-based—should be built in significant numbers to police, protect, and defend the extensive communications routes that are considered vital. Because of the primacy of this mission, space systems that perform purely offensive operations—those with negligible influence ensuring access to and use of CLOCs—are of secondary importance.[34]

Stuart Eves has written of a similar idea, using terminology *angels* and *demons*.[35] Due to space being considered a domain for military affairs where high-value assets may be considered targets for enemy action, there is a need to monitor and protect one's own satellites while also affecting negatively an adversary's satellite. Eves describes *angels* as small, autonomous, cooperative, co-orbiting guardian satellites operating near the target to provide "evaluation of the local space" around the larger host satellite.[36] In contrast, *demons* are small satellites designed to rendezvous with their targets to conduct intelligence-gathering operations, and in time of crisis, compromise the operations of the target.[37] Eves goes on to note that the functions of angels and demons should be considered inextricably linked because one seeks to counter the other. Others have also concluded on the need to protect high-value satellites. For example, Michael Nayak has argued for using cube satellites to protect and defend high-value satellites against an adversary's small satellite threat.[38] Nayak refers to these cube satellites performing the defensive space control mission as *guardians*.[39]

While technological capability is emerging to allow for angels and demons, the current reality is that space warfare has a mostly terrestrial look. Consequently, achieving a level of command of space and ensuring access to and use of CLOCs will entail employing terrestrial-based capabilities and services. This means terrestrially based space denial measure—like ground-based laser, anti-satellite weapons, and communications jammers—have their place as well in a practical strategy.

Money matters

When it comes to strategy, money matters. Bernard Brodie said this elegantly: "Strategy wears a dollar sign."[40] Strategy involves balancing ends and means; therefore, fiscal resources used to procure military forces and major defense programs dictate the means available to achieve political ends in war. And the amount of economic and fiscal resources that a country has will drive the political ends that can realistically be sought during conflict. Colin Gray underscores Bernard Brodie's observation above regarding the role of economics in shaping strategy by noting, "… politics is only the master of strategy it can afford."[41] For space strategy, this means the economic gain, including that achieved through commercial space activities, can be used to fund the development and employment of the means used to successfully win a war, especially during protracted conflicts or those where a decisive victory is elusive.

Historical experience from conflicts within the land, sea, and air domains supports the idea of affecting negatively an adversary's commerce and trade. Thucydides writes of the campaigns between Athens and Sparta, where commercial allies and partners were targeted, and the adversary's crops and harvests were destroyed. Germany's unrestricted submarine warfare during the World Wars sought to sink commercial shipping that benefited the Allies. The Allied air campaign against Germany in World War II targeted industrial and commercial facilities used to support German warfighting capability. The maritime strategist Julian Corbett wrote about this idea. He saw interfering with the enemy's trade as not only a means of exerting economic pressure, but also serving as a means of overcoming the enemy's "power of resistance."[42] The economy of a state is an important factor in sustaining a protracted war, and Corbett observed, "All things being equal, it is the longer purse that wins."[43]

Consistent with warfare within the other domains, space strategy will frequently need to affect negatively an adversary's space-related commerce and trade, to lessen the adversary's long-term warfighting ability. The intent of doing so is to bring the war to a decisive conclusion sooner. Later chapters—particularly Chapters 5, 6, and 8—will address economic and commercial measures in greater detail.

Offensive and defensive strategies

Today, attacking on-orbit capabilities is seen as potentially offering a means of crippling a state's conventional power projection and imposing significant costs, whether in dollars, lives, or political capital. Many strategists and policy-makers have concluded that because satellites and other space systems are seen as being exposed to attack—with little way to defend them—the offense is the stronger form of warfare in space. For example, it has been argued, "The offense may appear to be the stronger form of war in space, given the absence of terrain obstacles, the relative paucity of capital assets (and target), and the global consequences of military success or failure."[44] Given the current state of technology and how space-based technologies are presently employed, it is easy to see how such a conclusion could be reached. Space-based systems certainly appear to be vulnerable targets, because satellites in orbit follow mostly predictable paths with few places to hide or methods to defend themselves. Despite a realization of the vulnerability of satellites in orbit today and many individuals' views to the contrary, a careful study of time-tested theory and principles of war underscores that the defense is indeed the stronger form of warfare in space.

The purpose of space strategy is to ensure access to and use of CLOCs during times of either peace or war, and offensive and defensive strategies must support this objective. When the use of force is warranted, and decisive action is called for, space strategy must address the best method of achieving political and national ends using the means available. Sun Tzu noted the separate advantages of the offense and defense, in observing, "Invincibility lies in the defence; the possibility of victory in the attack."[45] Both offensive and defensive strategies have inherent strengths and weaknesses; therefore, the strategist must integrate them both into an overall strategic plan.

Offensive strategy

> There is only one means in war: combat.
>
> Clausewitz[46]

In Clausewitz's framework of a general theory of war, offensive strategy is aimed at acquiring or wrestling something from the enemy, and he noted that the destruction of the enemy's forces always offers the highest probability of victory. Therefore, the offense has a positive objective. According to Clausewitz, an offensive strategy is a preferred instrument for the stronger power.[47] Offensive actions can magnify the attacker's advantages gained through the strength and energy that comes from initiating an attack. Clausewitz acknowledges that initiative will initially favor the attacker because an attacker determines the time and place of initiating conflict. Nonetheless, the Prussian strategist advises that the offensive must not be confused with initiative itself because it is possible to gain the initiative during a counterattack, which is one of the central purposes of the defense.

There are other advantages to offensive strategy and actions. The psychological effect of a successful offensive attack can demoralize an enemy, thus contributing to the success of follow-on offensive operations. Successful offensive operations can beget more successful operations. Additionally, conducting offensive operations may make the adversary react to those actions, instead of carrying out his originally intended military strategy to achieve political ends.

Early airpower theory embraces the perceived advantages of offensive strategy and action. In the 1920s, Italian Air Marshal Giulio Douhet advocated that aircraft are the solution to strategic and tactical stalemates, and all future wars can be won from the air.[48] Douhet thought the aircraft's strategic advantage lay in its inherent offensive characteristics, which are freedom of maneuver and speed.[49] Moreover, any defense will be ineffective against aircraft, and "no fortification can possibly offset these [aeroplanes], which can strike mortal blows into the heart of the enemy with lightning speed."[50] For Douhet, the formula for victory included gaining command of the air and then attacking the enemy's industrial and commercial facilities, critical transportation centers and routes, as well as designated civilian population areas.[51]

Other air power theorists also saw the advantages of offensive action through aircraft. In 1921, William "Billy" Mitchell stated, "[A]s air covers the whole world, aircraft are able to go anywhere on the planet ... [and] have set aside all ideas of frontiers."[52] Mitchell assessed that some air operations—such as strategic bombing—can achieve independent results, thereby winning wars by destroying the enemy's war making capability and will to fight.[53] Additionally, he believed that the first battles during future wars would be air battles, and that the state that wins them was "practically certain to win the whole war."[54] In 1988 and using Clausewitz's center of gravity concept for inspiration, John Warden theorized that air power possesses the capacity to achieve victory with maximum effectiveness and minimum cost.[55] Warden's theory is visualized as a series of concentric rings relating to society. The most important of these rings is at the center, representing enemy leadership. Because of the leaders' strategic level decision-making ability, offensive efforts should be directed there, and air power is ideally suited for that mission.[56] J.C. Wylie summarized his view of air power theory in writing:

> The air theory has as an essential, though tacit, assumption the premise that control of a people can in fact be exercised by imposition (or threat of imposition) of some kind of physical destruction, and that, furthermore, this can be imposed from the air.[57]

Taken together, the theories of Douhet, Mitchell and Warden advocate using air forces to achieve strategic effects through offensive strategy and action, including aerial bombing.[58]

In looking to determine if U.S. joint service space doctrine incorporates this view of offensive strategy, the unfortunate answer is it does an abysmal job. The closest concept analogous to offensive strategy in U.S. joint doctrine is *offensive* space control.[59] According to Joint Publication 3–14, *Space Operations*, offensive space control is said to include "offensive operations conducted for space negation, where negation involves measures to deceive, disrupt, degrade, deny, or destroy space systems and services."[60] Despite an attempt to provide strategy through alliteration, the concept of offensive space control fails to incorporate that offensive strategy is aimed at acquiring or wrestling something from the enemy. By using terms like *deceive* and *disrupt*, the true nature of offensive strategy—which includes combat and violence—is hidden.

Defensive strategy

> Defense is the stronger form of waging war.
>
> Clausewitz[61]

In contrast, the objective of a defensive strategy is to preserve one's forces, assets, or capabilities or to prevent the enemy from acquiring something or achieving a political objective.[62] Consequently, the defense has a negative objective. Clausewitz asserts the superior strength of the defense, despite the considerable advantages the attacker might enjoy during the opening phases of hostilities. He contends, "So in order to state the relationship precisely, we must say that *the defensive form of warfare is intrinsically stronger than the offensive*."[63] Clausewitz concluded that a defensive strategy should be assumed when one is weaker than the adversary, and a defensive strategy should be abandoned once one is able to pursue an offensive strategy.[64] A defensive strategy includes an attitude of alert expectation that waits for the moment when the enemy exposes himself, in order to launch a successful counter-attack.[65]

At the operational and tactical levels of warfare, Clausewitz observes that defensive strength also derives from preparations taken prior to conflict. In describing how to implement a sound defensive strategy, he states, "Part of strategic success lies in timely preparation for a tactical victory.... The rest of strategic success lies in the exploitation of victory won."[66] Additionally in his *Principles of War*, he notes that the "defense is nothing more than a means by which to attack the enemy most advantageously, in a terrain chosen in advance, where we have drawn up our troops and have arranged things to our advantage."[67] Consequently, a defensive strategy includes requisite preparatory actions ahead of hostilities, to exploit situational advantages created during the course of war.

What does this mean for the development of space strategy? It means defensive strategy is the stronger form of warfare in space, but assuming a defensive posture requires making adequate preparations. These defensive preparations may include several actions taken before and after hostilities commence, such as knowledge of the adversary's forces and disposition, capability to gain indications and warnings of initial hostilities, and understanding when and under what conditions a counterattack will be most successful. These preparatory actions presuppose significant space situational awareness, along with ample systems and space professionals for collecting and analyzing intelligence and developing an understanding of the expected threat. Additionally, taking the requisite preparations may include incorporating self-protection countermeasures, hardening spacecraft, diversification, distribution, satellite maneuverability, and multi-domain combat solutions. These ideas are in line with many of the tenets of space mission assurance and resilience.[68]

Failure to recognize the importance of these preparations and to place critical space-based systems in unprotected, highly vulnerable locations is an invitation for attack. Failure to make adequate preparations ahead of time will negate the inherent advantages of a defensive strategy. Space strategist M.V. Smith has hinted at this exact situation in saying, "It is often said that the defense is the stronger form of warfare. This is not true in space—today."[69] Certainly at present, it seems that many space powers have failed to make the necessary investments of incorporating those measures providing protection and defensive capability, regardless whether terrestrially or space-based. Once these essential pre-hostility actions and preparations are taken, the practical execution of space strategy will more fully embrace the principle that the defense is the stronger form of war in space.

It is worthwhile comparing the Clausewitzian approach to defensive strategy to that of U.S. joint service doctrine—Joint Publication 3–14, *Space Operations*. In the publication, defensive strategy is predominantly described under *defensive space control*, which is said to be "active and passive measures taken to protect friendly space capabilities from attack, interference, and unintentional hazards."[70] While joint service doctrine notes the need to preserve and protect capability, it does not describe the defense being the stronger form of war or that the defense should include making necessary preparations against potential attack.

Thucydides observed, "War is a matter not so much of arms as of money," and this is especially true for implementing a sound defensive strategy.[71] There is a constant give and take between the need to develop and build the most defendable space systems and being able to afford them. The most defendable system may cost more to develop and weigh more, thereby costing more to place in orbit. Strategists and warfighters will need to consider the benefits of defensive approaches, along with associated time and fiscal procurement costs, when finally deciding upon the best approach. It may be the case that less physical protection (e.g., hardening) is needed in favor of increasing the numbers of assets to support distribution and diversification measures.

Offense and defense are interdependent

> If defense is the stronger form of war, yet has a negative object, it follows that it should be used only so long as weakness compels, and be abandoned as soon as we are strong enough to pursue a positive object.
>
> Clausewitz[72]

For space strategy—or any strategy for that matter—it is best to consider the defense working with the offense to win the war and achieve political ends.

Offensive and defensive strategies are each mutually dependent and so inter-twined that one is not ultimately successful without the other. Defensive operations protect the very lines of communication that make offensive operations possible. If requisite preparations are taken and operations are implemented properly, defensive strategies frequently require fewer forces and assets than offensive strategies; therefore, defensive operations in some regions facilitate the concentration of military forces or effects to support offensive operations in other regions.

Regarding the interdependency of the offense and defense, Colin Gray observes, "Operational offense married to tactical defense can be the most lethal of combinations for the enemy."[73] However, Gray cautions the strategist in noting that although there are times when a good offense is a good defense and vice versa, overly enthusiastic pursuit of either style of warfare tends to become self-defeating.[74] While it is generally true that war can only be won on the offensive, it cannot be won by the offensive alone. Undue devotion to either the offense or defense during warfare poses severe problems of balance and thus can undermine the effectiveness of strategic operations.[75]

At times the line between offensive and defensive strategy becomes blurred because of the inherent interdependency. Julian Corbett argued that at the heart of defensive strategy is the counter-attack.[76] Although a counter-attack is indeed a defensive response, if the counter-attack becomes a sustained operation, it can become offensive in scope and intent. Similarly, an offensive operation that encounters resolute resistance by an enemy may experience an "operational pause," where additional logistical support is brought into the theatre of operations. In this situation, the scope and intent become defensive until an offensive strategy can be resumed.

Jointness

Operations in space are interdependent with operations on land, at sea, in the air, and in cyberspace. A state's grand strategy goals should be reflected in its national policy and strategy. If its efforts are properly marshalled, all sub-strategies—such as land, sea, air, space, and cyberspace—should work towards those goals to achieve political ends. Past conflicts show that hostilities are rarely concluded in a lasting way until land forces are brought to bear against the adversary, or at the least the threat of bringing them to bear is perceived as being real and imminent. In modern warfare, to get to the point where land forces can threaten an adversary may require the combined efforts of the other relevant warfighting domains.

The thought that the military branches should work together is described by Charles E. Callwell (1859–1928), a British field artillery officer who wrote on maritime strategy. His work about maritime strategy and amphibious warfare, *Military Operations and Maritime Preponderance*, describes the interdependence of land and maritime warfare that still has relevance today.[77] Callwell wrote about the application of military force from the sea, and his treatment of land

and sea forces is one of the first and most significant discussions of what has become known since as "joint" warfare.[78] Callwell commented on how naval preponderance and land warfare are mutually dependent and that there is an "intimate connection" between command of the sea and control of the shores.[79] He thought that a great land campaign that is reliant on the sea is impossible without maritime preponderance.[80] In order for naval and land forces to achieve the greatest effect, they must act in harmony and in constant coordination at all times. Callwell believed that "United we stand, divided we fall" is an applicable motto expressing the need for a maritime nation to have the closest of cooperation between its land and sea forces.[81] While acknowledging the downside of using strategic analogies, a reasonable lesson for space strategy is that space forces must operate in concert with other military forces to achieve the greatest effect.

Current U.S. joint doctrine embraces the idea that space-based capabilities and services can promote jointness of the military services. The 2018 joint doctrine, *Space Operations*, notes that space operations support air, land, maritime, and cyberspace fires through positioning, navigation, and timing and satellite communications capabilities, which in turn increase operational tempo, dispersion, and concentration of forces.[82]

Strategy for the less capable

Based upon the separate advantages of offensive and defense strategies, along with an understanding of their interdependence, a framework for how those who are less capable in space can be better understood. Although a less capable space force is unlikely to win a major and decisive space engagement against a superior opponent, less capable forces can still contest a more capable power's command of space and, in doing so, achieve limited political objectives. Methods to contest the command of another include both non-military and military actions. Through this method, a lesser power may attempt to bolster its power and influence, while diminishing the instruments of power of a superior adversary. This section summarizes many of the potential strategies and actions by less capable space powers, but these ideas are examined in depth in Chapters 7 and 8.

Regarding the use of non-military actions, it is reasonable to presume that a less capable space power will attempt to use the most effective instruments at its disposal, which may include non-military means such as diplomacy, economic, and informational instruments of power. First, those less capable in space can improve diplomatic influence by establishing a notable presence in space and then proposing international treaties, agreements, principles, or resolutions that advance their interests on relevant issues. Although not an absolute prerequisite, those with the most presence in outer space and the most space-based activities will have the greatest chance of shaping international laws, regulations, and norms of behavior. Second, those less capable in space can use economic measures to contest command of space and achieve modest result. A lesser power that provides a unique commercial or business service can threaten to withhold its space-based service in the hopes of negotiating better terms on some contentious issue. Third, informational actions can be used to achieve positive results. Indeed, informational methods are frequently the simplest means with which to contest a superior space power. By conducting a sustained campaign to promote news or viewpoints that advances its long-term strategy, a lesser power, over time, may change what is perceived or considered as true by others.

When considering the military instrument of nation power, a less-capable space power can contest the command of space to achieve limited political objectives. Because a lesser force is by definition less capable militarily relative to a stronger force, the lesser most often will need to gain local or temporary command in areas where the stronger force is not, thereby contesting the command exercised by the stronger force. By attacking where or when the other is not strong, temporary command can be gained and exercised to achieve political ends.

Another effective method by which a lesser space force can contest command is the "force in being" concept, which is based upon the "fleet in being" concept of maritime strategy.[83] While avoiding a decisive military engagement against a stronger space power, a less capable space force should be kept "in being" through active utilization and operation to achieve limited political ends until the situation improves in its favor. By avoiding large-scale engagements with a superior space force, a lesser one can conduct minor, non-escalatory frustrating and harassing operations along CLOCs or against space-related activities, thus preventing a more capable power from gaining command of space that is either general or persistent. The force in being can include both terrestrial and space-based capabilities, including direct ascent anti-satellite weapons and on-orbit radio frequency interference.

Dispersal and concentration

Because space is vast—with expansive CLOCs—and a state's available resources are finite, space strategy must address the ways and means of employing and distributing assets to achieve political ends. This requires implementing an approach that combines the ideas of dispersal and concentrations. Alfred Thayer Mahan explained this complex idea through a simile: "Such is concentration reasonably understood—not huddled together like a drove of sheep, but distributed with a regard to a common purpose, and linked together by the effectual energy of a single will."[84] Likewise, Sun Tzu noted the importance of this idea when writing, "Move when it is advantageous and create changes in the situation by dispersal and concentration of forces."[85]

The first part—dispersal—considers the distribution of forces, assets, and effects wherever vital space interests are located along CLOCs. Dispersing forces, assets, and effects to the widest extent practical potentially bestows on a space power those benefits coming from military presence together with the potential coercive effects and increased uncertainty for the adversary that results.

By dispersing space systems within a certain region or along CLOCs, a potential adversary maybe uncertain as to the force's disposition and intended function.

The need to disperse assets and systems is somewhat reflected—although not completely—in the concepts of mission assurance and resilience, as defined by U.S. defense department space domain mission assurance frameworks. Part of *resilience* is said to include distribution and diversification.[86] *Distribution* is utilizing a number of nodes, all working together, to perform the same mission or functions as a single node. An example of distribution is said to be the Global Positioning System, where no individual satellite or ground monitoring site is fundamental to assuring positioning, navigation, and timing in any one specific location. In contrast, *diversification* is defined as contributing to the same mission in multiple ways, using different platforms, different orbits, or systems and capabilities of commercial, civil, or international partners. Diversification includes those systems or architectures that are flexible or adaptable for supporting a variety of missions or functions. Admittedly, distribution and diversification, as defined in U.S. defense department mission assurance framework documents, does not capture the idea of dispersal in totality, because dispersal is not described as action relative to CLOCs.

The other half of the concept, *concentration*, refers to focusing firepower, assets, or other desired effects to defeat an adversary, defend against its attack, or neutralize the threat the enemy poses. The principle of concentration, therefore, tells the strategist that when offensive actions are imminent or necessary, the greatest force that is practical should be concentrated against the enemy.[87] This firepower or neutralizing effect can originate from military systems that communicate force into, from, and through space. Examples may include ground-based lasers, air-launched anti-satellite weapons, or space-based weapons.

Some strategists may take exception to the idea of dispersal and concentration, especially with the concentration of systems or forces. Critics may note their objection in the apparent maneuvering limitations of systems on-orbit with respect to Keplerian physics. Satellites are not ships, which can readily change course. Satellites must follow predetermined paths, and consequently, there is no historical experience that substantiates that the concept of concentration applies in space. Such sentiments are a half-truth, at best.

Yes, the laws of physics are important in space operations and the development of space strategy—as they are in all domains—but satellites in near-Earth orbit or operating in other regions are not hapless platforms beholden to orbital mechanics. Recent advancements in rendezvous and proximity operations by the United States, Russia, and China illustrate this. Just as ships were limited to trade winds routes during the Age of Sail, technology and propulsive capabilities advanced allowing ships to maneuver across their domain of operation. This process will hold for operations in space as well. Additionally, concentration in the context of space strategy will apply to non-kinetic effects, including communication jamming, laser interference, and cyber actions against space-related infrastructure and networks.

Within U.S. joint doctrine, the idea of concentration is somewhat covered. The 2013 Joint Publication 3–14, *Joint Doctrine for Space Operations*, listed principles of war as related to space warfare. Using Clausewitz's principle of mass, the joint document notes that "The purpose of mass is to concentrate the effects of combat power at the most advantageous place and time to produce decisive results."[88] The publication describes that it is critical for commanders to integrate and synchronize supporting space forces, so that the concentration of combat power at the proper time and place can be most effective, thereby conserving available resources, minimizing impact on non-adversaries, and maximizing the effect on the adversary.[89]

The concept of dispersal and concentration is different from the principle of concentration within land warfare theory. Agreeing with Mahan's quote at the beginning of this section, Corbett thought naval concentration meant assembling "the utmost force at the right time and place," and it includes the capability to stop the concentrating process and rapidly shift the direction of naval forces.[90] So, despite the apparent benefits coming from the concentration of military force, concentration as a principle unto itself is incomplete. Dispersal and concentration needing to be linked together by Mahan's idea of the "effectual energy of a single will" is also reflected in the writing of Raoul Castex (1878–1968), who wrote on the interaction of land and sea on the army and navy. Castex observed that ships need to move "intelligently" to achieve strategic advantage, an idea he called *manoeuvre*.[91] Castex's idea of *manoeuvre* includes moving one's forces, resources, or capabilities to various areas that are more profitable for overall strategic results, usually by gaining some measure of success in secondary theaters that can be translated into support of the main theater of war.[92]

Ultimately, a space strategy that is practical in its execution requires a balance between dispersal and concentration. Space strategy necessitates that forces, assets, and effects should, in general, be dispersed to cover the widest possible area, yet retain the ability to concentrate force or effects rapidly when needed. Therefore, the application of either dispersal or concentration should be considered as one and the same and not as separate and discrete principles in themselves. Employing an integrated approach to dispersal and concentration will preserve the flexibility of protecting a state's most vital CLOCs while better engaging an adversary's "central mass" when and where needed.[93]

Blocking CLOCs

The strategy of space warfare must also determine where, when, and in what manner to block an adversary's use of CLOCs. *Blocking* is the act of disrupting, degrading, or denying an adversary's capability to use his CLOCs, thus minimizing the movement of spacecraft, equipment, materiel, supplies, personnel, military effects, data, or information.[94]

Drawing upon the historical experience of maritime strategy, blocking is best considered in two general categories: close and distant.[95] These two categories refer to where blocking is employed relative to hubs of activity or points of distribution. Close blocking equates to preventing the deployment, launch, or movement of space systems near hubs of activity. It also pertains to interfering with communications near uplinks, downlinks, and crosslinks. In contrast, a capable space power can employ distant blocking to force an adversary into action, by occupying or interfering with the distant and potentially common CLOCs.

Blocking CLOCs incorporates elements of both offensive and defensive strategy. Offensive strategy is used when political objectives necessitate taking or acquiring something from the enemy, and defensive strategy is used when political objectives necessitate preventing the enemy from achieving or gaining something. In blocking, the intent may include wresting lines of communication away from the enemy, thereby taking them for oneself. Thus, the intent in this case is more offensive in nature. In addition, lines of communication in space are often shared, and a state may initially enjoy equal access to the same lines of communication as an enemy state. In such cases, the purpose would not be to acquire access to these lines of communication, but it would only be to prevent the enemy from using of them. Therefore, when sharing CLOCs with the adversary, the blocking strategy is more defensive rather than offensive in nature.

Within U.S. joint service doctrine, the idea of blocking CLOCs finds its closest similarity to the term *space negation. Space negation* is said to involve:

> Active defensive and offensive measures to deceive, disrupt, degrade, deny, or destroy an adversary's space capabilities. Measures include actions against ground, data link, user, and/or space segment(s) to negate adversary's space systems, or to thwart hostile interference with or attacks on US/allied space systems.[96]

Furthermore, in examining the concepts included in *space negation,* the 2013 U.S. joint doctrine defines *deception* as those measures designed to mislead an adversary by manipulation, distortion, or falsification of evidence to induce the adversary to react in a manner prejudicial to their interests. *Disruption* includes those measures designed to temporarily impair specific targeted nodes of an adversary system, usually without physical damage to the space system. *Degradation* is composed of those measures designed to permanently impair—either partially or totally—the utility of targeted adversary systems, usually with physical damage. *Denial* activities are those designed to temporarily eliminate the utility of targeted adversary systems, usually without physical damage. Finally, *destruction* is defined as those measures designed to permanently eliminate the utility of targeted adversary systems.[97] While the definition of *space negation* within joint service doctrine does not explicitly include the idea of affecting negatively an adversary's access to and use of CLOCs, it is somewhat implied because the concept is included under the section on *space control,* whose definition does include freedom of action for friendly forces and defeating an adversary's capabilities.

Positions

> The second rule is to concentrate our power as much as possible against that section where the chief blows are to be delivered and to incur disadvantages elsewhere, so that our chances of success may increase at the decisive point. This will compensate for all other disadvantages.
>
> Clausewitz, Principles of War.[98]

As with the other warfare domains, there are advantageous and valuable positions in space warfare. These positions are locations that impart some relative advantage or are more conducive for achieving a strategic effect when operating there. Such positions include those for the movement of physical assets like space vehicles, equipment, materiel, supplies, and personnel, along with those pertaining to non-physical communications like some weapons effects and electromagnetic transmissions. Sun Tzu writes on the advantages of position during war by writing, "Therefore the skillful commander takes up a position in which he cannot be defeated and misses no opportunity to master the enemy."[99] In Sun Tzu's example, however, the only preparation considered is holding better geography. Perhaps one of the more well-known concepts related to positions is Clausewitz's center of gravity concept. Regarding the idea, he said "To achieve victory we must mass our forces at the hub of all power and movement, the enemy's center of gravity."[100] In this context, the center of gravity represents important capabilities for the enemy or a source of strength, which if attacked, can lead to an expedited victory.

A variation of Clausewitz's thinking is John Warden's five rings from air-power theory, described in his article "The Enemy as a System."[101] In commenting on the utility of the Clausewitzian centers of gravity, Warden wrote:

> The concept of centers of gravity is simple in concept but difficult in execution because of the likelihood that more than one center will exist at any time and that each center will have an effect of some kind on the others.

According to Warden, a better approach is a five-ring model, where each level of system or "ring" is considered one of the enemy's centers of gravity. The idea behind Warden's five rings approach was to attack each of the rings to paralyze the enemy, an objective known as *strategic paralysis*.[102] In describing his approach, Warden says:

> If we are going to think strategically, we must think of the enemy as a system composed of numerous subsystems. Thinking of the enemy in terms of a system gives us a much better chance of forcing or inducing him to make our objectives his objectives and doing so with the minimum effort and the maximum chance of success.[103]

Warden's five rings comprise: leadership; organic essentials (e.g. electrical power); infrastructure; population; and fielded military forces.[104]

In *Strategy of the World War*, Vice-Admiral Wolfgang Wegener (1875–1956), a German career naval officer, wrote on the importance of positions in war.[105] He argued that having a naval fleet was meaningless if it did not control a strategic position that affected the enemy's commerce and trade.[106] Therefore, contrary to the thinking of the German naval leadership during the war, the German fleet needed to threaten Great Britain's strategic positions, to force a decisive naval battle. Consequently, the role of geography should have been considered when developing German naval strategy, and strategists should not have focused just on tactical-level offensive fleet actions. As a result, the German naval strategy was doomed to failure from its inception. According to Wegener, the Battle of Jutland meant nothing in the end because it did not improve Germany's control of strategic positions or affect British maritime commerce. In contrast, he thought that the early successes of the German submarines during the World War I were the correct application of sea power, because they attempted to influence negatively British commerce and access to vitally important sea lines of communication.[107]

Historical experience from land, naval, and air warfare underscores that positions can be exploited for military advantage to achieve strategic effect. The lesson for space strategy is that by exploiting advantageous and valuable positions for strategic effect, a space force can potentially restrict the movement of the enemy's forces, processes, or available information—thereby improving the conditions for military operations. If correctly exploited, these positions can restrict an adversary's access to and use of CLOCs, thus improving one's overall conditions for military success at the tactical, operational, or strategic levels of war. When considering positions, they can be described broadly as choke points, high-value positions, and high ground.[108]

Choke points

As with crossroads, straits, and airfields in land, naval, and air operations, space has locations and regions where assets and communications tend to converge or have focal points. These choke points include ground-based uplinks or celestially based downlinks and crosslinks used to transmit and receive data and information. Choke points can also include frequency spectrum bands that are more desirable or more predominately used, even if shared between adversaries. Additionally, choke points can pertain to physical assets, to include significant concentrations of spacecraft or systems, thereby enabling the restriction of an adversary's access to and use of space. These locations may include space launch locations and space stations or bases that are used for scientific, commercial, logistical, or military enterprise.

High-value positions

Although choke points are locations or regions that can be exploited for military advantage, other positions may also hold strategic value as well. These high-value positions are commonly space-based systems performing valuable or unique services, whether for economic, informational, or military uses. Albeit high-value positions may also represent choke points for space communications, this is not necessarily always the case. Examples of high-value positions include satellites that are part of positioning, navigation, and timing constellations, to include: American GPS, Chinese Beidou, Russian Global Navigation Satellite System (GLONASS), and European Galileo. Because these satellite constellations are designed to have about two dozen satellites—in addition to orbital spares—as part of each constellation, the loss of a single satellite does not result in a critical loss of capability. Consequently, any single satellite does not represent a choke point in itself. Although these positioning, navigation, and timing satellites are not considered choke points, they are nonetheless considered high-value positions with strategic value. If an adversary can destroy or degrade enough positioning satellites to make the constellation's services ineffective, financial and commercial sectors that rely on the positioning and timing information will be impacted negatively, to the point of potentially creating a national crisis. For this reason, military space strategy is concerned with protecting a state's high-value positions and affecting negatively those of an adversary.

High ground

Outer space, especially near-Earth orbits, has been touted as the "ultimate high ground" for some time.[109] As with the terrain used by artillery overlooking enemy formations or the high altitude flown by bomber aircraft, assets in space have a superior view of the Earth and therefore may enjoy a strategic advantage. This "high ground" is thought to give one the ability to employ weapons more effectively and conduct surveillance against the enemy below. Intelligence, surveillance, and reconnaissance satellites have long used the extensive field of view that comes from having systems in orbit around Earth. Depending on the type of orbit, sensor capabilities, and its height above the Earth, surveillance satellites can be optimized to observe specific terrestrial features or geographic areas.

Everett Dolman describes the positional advantage relative to other space objects as being higher in the "gravity well."[110] Dolman states that an object higher in the gravity well has an advantage over a lower object. Because of the Earth's gravitational pull and the potential energy coming from operating high above the Earth, this energy can be imparted to kinetic energy warheads, thereby providing more destructive firepower.

The advantage realized by high ground is, however, a relative measure. This advantage does not occur between two space-based weapons with similar capabilities. When compared to many ground systems, those in orbit have distinct advantages in military utility. However, against a comparable space-based asset, the benefit may be non-existent.

Strategy as a practical matter

The nature of war is enduring, even though its character changes. Space warfare will have a character different than that of land, sea, and air. Also, the character of war in space and the requisite military approaches will change during the passage of time and from conflict to conflict. Nonetheless, the previous areas of discussion highlight some of the enduring principles and concepts that the policy-maker, strategist, and warfighter should consider when looking to protect national interests in space. Bernard Brodie writes that strategy "is nothing if not pragmatic," and consequently, the implementation of space strategy is a practical matter.[111]

Yet, any general theory of space strategy will have limitations. Clausewitz made it clear that theory and strategic principles are never a substitute for good judgment and experience.[112] He believed that a theory-based strategy helps to determine a coherent plan for war, but should not be blindly trusted in action.[113] Individual thought and common sense should remain masters, providing guidance when the situation is uncertain.[114] Furthermore, the practical value of theory—including the theory of space strategy—is its ability to assist in acquiring a broad outlook, whereby the factors of a sudden predicament may be rapidly ascertained. In the end, strategic theory must not only be able to make sense of what has occurred in the past, but also be able to provide some basis for considering plausible and likely events in the future.

Finally, much of the problem regarding the development of space strategy is that our historical experience has been mostly concerned with near Earth—and more recently cislunar—activities and national interests. Yet, space is so much more than that. A general theory of space strategy should be timeless in describing the nature of war in space, even as activities and national interests move outward from Earth. It is thought that the areas and topics covered in this chapter will remain relevant in the future.

Notes

1 Carl von Clausewitz, *Vom Kriege*, erster Band (Berlin: Ferdinand Dümmler, 1832), 92.
2 John B. Sheldon and Colin S. Gray, "Theory Ascendant? Spacepower and the Challenge of Strategic Theory," in *Toward a Theory of Spacepower: Selected Essays*, eds. Charles D. Lutes and Peter L. Hays, with Vincent A. Mazo, Lisa M. Yambrick, and M. Elaine Bunn (Washington, DC: National Defense University Press, 2011), 12.
3 Ibid.
4 Carl von Clausewitz, *On War*, trans. and eds. Michael Howard and Peter Paret (Princeton, NJ: Princeton University Press, 1989), 140.
5 John J. Klein, *Space Warfare: Strategy, Principles and Policy* (Abingdon: Routledge, 2006). The space strategy framework described in this chapter comes, in part, from this book and Sir Julian Corbett's book *Some Principles of Maritime Strategy*. Julian S. Corbett, *Some Principles of Maritime Strategy* (London: Longmans, Green and Co., 1911; reprint, Annapolis, MD: Naval Institute Press, 1988).
6 Including Brent Ziarnick, Bleddyn Bowen, and this author. Brent Ziarnick, *Developing National Power in Space: A Theoretical Model* (Jefferson, NC: McFarland,

2015); Bleddyn E. Bowen, *Spacepower and Space Warfare: The Continuation of Terran Politics by Other Means* (PhD Thesis, Aberystwyth University, 2015); Klein, *Space Warfare.*

7   Giulio Douhet, *The Command of the Air*, trans. Dino Ferrari (1921 and 1927; New York: Coward-McCann, 1942; reprint, Washington, DC: Air Force Museums and History Program, 1998), 25.

8   William Mitchell, *Winged Defense: The Development and Possibilities of Modern Air Power—Economic and Military* (1925; reprint, New York: Dover Publications, 1988), xiii.

9   Alfred T. Mahan, *The Problem of Asia and Its Effect Upon International Policies* (Boston: Little, Brown, and Company, 1900), 125.

10 Jiang Lianju and Wang Liwen, eds., *Textbook for the Study of Space Operations* (Beijing: Military Science Publishing House, 2013), 6; quoted in Kevin L. Pollpeter, Michael S. Chase, and Eric Heginbotham, *The Creation of the PLA Strategic Support Force and Its Implications for Chinese Military Space Operations* (Santa Monica, CA: RAND Corporation, 2017), 7.

11 The White House, *The National Security Strategy of the United States of America* (December 2017), www.whitehouse.gov/wp-content/uploads/2017/12/NSS-Final-12-18-2017-0905.pdf

12 European Commission, *Space Strategy for Europe*, Communication from the Commission to the European Parliament, the Council, the European Economic and Social Committee and the Committee of the Regions (October 26, 2016), 8.

13 Office of the Assistant Secretary of Defense for Homeland Defense and Global Security, *Space Domain Mission Assurance: A Resilience Taxonomy* (September 2015), http://policy.defense.gov/Portals/11/Space%20Policy/ResilienceTaxonomy-WhitePaperFinal.pdf?ver=2016-12-27-131828-623

14 Ibid., 2.

15 Ibid., 6.

16 Ibid.

17 Klein, *Space Warfare*, 51, 155

18 Ibid.

19 Ibid., 60.

20 Bowen, *Spacepower and Space Warfare*, ii.

21 Julian S. Corbett, *Some Principles of Maritime Strategy*, 91.

22 Joint Chiefs of Staff, *Space Operations*, Joint Publication 3–14 (April 10, 2018), I–3, www.jcs.mil/Portals/36/Documents/Doctrine/pubs/jp3_14.pdf

23 J.C. Wylie, *Military Strategy: A General Theory of Power Control*, with introduction by John B. Hattendorf (New Brunswick, NJ: Rutgers University Press, 1967; reprint, Annapolis, MD: Naval Institute Press, 1989), 77–78.

24 Colin S. Gray, *Airpower for Strategic Effect* (Maxwell Air Force Base, AL: Air University Press, 2012), 278.

25 Ibid., 282.

26 Klein, *Space Warfare*, 104–105.

27 Corbett, *Some Principles of Maritime Strategy*, 318.

28 People's Liberation Army Academy of Military Science, Military Strategy Studies Department, *Science of Military Strategy* (Beijing: Military Science Press, December 2013), 182; from Pollpeter, Chase, and Heginbotham, *The Creation of the PLA Strategic Support Force*, 9.

29 Klein, *Space Warfare*, 100–106.

30 Ibid.

31 Francis Bacon, "Of the True Greatness of Kingdoms and Estates," in *The Essays* (1601; reprint, Adelaide: The University of Adelaide, 2014).

32 Joseph S. Nye, Jr., *Softpower: The Means to Success in World Politics* (New York: Public Affairs, 2005).

33 Sun Tzu, *The Art of War*, trans. Samuel B. Griffith (Oxford: Oxford University Press, 1963), 77.

34 Although considered secondary, these other functions that are performed may, in fact, be essential to winning the war.

35 Stuart Eves, "Angels and Demons: Cooperative and Non-cooperative Formation Flying with Small Satellites" (presentation, Surrey Satellite Technology Limited, London, University of London, 2008), 2.

36 Ibid.

37 Ibid.

38 Michael Nayak, "Deterring Aggressive Space Actions with Cube Satellite Proximity Operations: A New Frontier in Defensive Space Control," *Air and Space Power Journal* vol. 31 no. 4 (Winter 2017), 92–102.

39 Ibid.

40 Bernard Brodie, *Strategy in the Missile Age* (Santa Monica, CA: RAND Corporation, 1959), 358.

41 Colin S. Gray, *Fighting Talk: Forty Maxims on War, Peace, and Strategy* (Westport, CT: Greenwood Publishing, 2007), 138.

42 Corbett, *Some Principles of Maritime Strategy*, 102.

43 Ibid., 101.

44 Colin S. Gray, *Weapons Don't Make War: Policy, Strategy, and Military Technology* (Lawrence, KS: University Press of Kansas, 1993), 15.

45 Sun Tzu, *The Art of War*, 85.

46 Clausewitz, *Vom Kriege*, erster Band, 47.

47 Clausewitz, *On War*, 97, 358; Corbett, *Some Principles of Maritime Strategy*, 31–33, 310–311. Corbett viewed the offense being the more "effective" form of warfare.

48 Douhet, *The Command of the Air*, 15–29.

49 Ibid., 15.

50 Ibid.

51 Ibid., 20.

52 Mitchell, *Winged Defense*, 4.

53 Mark A. Clodfelter, "Molding Airpower Convictions: Development and Legacy of William Mitchell's Strategic Thought," in *The Paths of Heaven: The Evolution of Airpower Theory*, ed. Phillip S. Meilinger (Maxwell Air Force Base, AL: Air University Press, 1997), 79.

54 William Mitchell, *Our Air Force: The Keystone of National Defense* (New York: E.P. Dutton and Co, 1921), xix.

55 David S. Fadok, "John Boyd and John Warden: Airpower's Quest for Strategic Paralysis," in *The Paths of Heaven, The Paths of Heaven: The Evolution of Airpower Theory*, ed. Phillip S. Meilinger (Maxwell Air Force Base, AL: Air University Press, 1997), 371.

56 John Warden, "The Enemy as a System," *Airpower Journal* vol. 9 no. 1 (Spring 1995): 41–55. www.airuniversity.af.mil/Portals/10/ASPJ/journals/Volume-09_Issue-1-Se/1995_Vol.9_No1.pdf

57 Wylie, *Military Strategy*, 63.

58 Douhet, *The Command of the Air*; Mitchell, *Winged Defense*; Clodfelter, "Molding Airpower Convictions," 79, 95.

59 Joint Chiefs of Staff, *Space Operations* (April 10, 2018), II–2.

60 Ibid.

61 Carl von Clausewitz, *Vom Kriege*, zweiter Band (Berlin: Ferdinand Dümmler, 1832), 147.

62 Clausewitz, *On War*, 357–358.

63 Ibid., emphasis original.

64 Ibid.

65 Ibid.

66 Ibid., 363.
67 Carl von Clausewitz, *Principles of War*, trans. Hans Gatzke (1812; The Military Service Publishing Company, 1942; Clausewitz.com, 2016), 4. www.clausewitz.com/readings/Principles/Clausewitz-PrinciplesOfWar-ClausewitzCom.pdf
68 Office of the Assistant Secretary of Defense for Homeland Defense and Global Security, *Space Domain Mission Assurance: A Resilience Taxonomy*.
69 M.V. Smith, "Spacepower and the Strategist," in *Strategy: Context and Adaption from Archidamus to Airpower*, eds. Richard J. Bailey Jr., James W. Forsyth Jr., and Mark O. Yeisley (Annapolis, MD: Naval Institute Press, 2016), 171.
70 Joint Chiefs of Staff, *Space Operations* (April 10, 2018) GL-5.
71 Robert B. Strassler, *The Landmark Thucydides: A Comprehensive Guide to the Peloponnesian War* (New York: Free Press, 1996), 46.
72 Clausewitz, *Vom Kriege*, zweiter Band, 146.
73 Gray, *Airpower for Strategic Effect*, 293.
74 Gray, *Weapons Don't Make War*, 23.
75 Ibid.
76 Corbett, *Some Principles of Maritime Strategy*, 32.
77 Charles E. Callwell, *Military Operations and Maritime Preponderance: Their Relations and Interdependence*, with introduction by Colin S. Gray (London: William Blackwood and Sons, 1905; reprint, Annapolis, MD: Naval Institute Press, 1996).
78 Colin S. Gray, introduction to *Military Operations and Maritime Preponderance: Their Relations and Interdependence*, Charles E. Callwell (London: William Blackwood and Sons, 1905; reprint, Annapolis, MD: Naval Institute Press, 1996), lxi.
79 Callwell, *Military Operations*, 443.
80 Ibid., 63.
81 Ibid.
82 Joint Chiefs of Staff, *Space Operations* (April 10, 2018), II–6.
83 Corbett, *Some Principles of Maritime Strategy*, 166.
84 Alfred Thayer Mahan, *Sea Power in its Relation to the War of 1812: Volume I* (Boston, MA: Little & Brown, 1905), 316.
85 Sun Tzu, *The Art of War*, 106.
86 Office of the Assistant Secretary of Defense for Homeland Defense and Global Security, *Space Domain Mission Assurance: A Resilience Taxonomy*, 6–7.
87 From an understanding of the Law of Armed Conflict, it is realized that the amount of force that is "practical" involves taking into consideration the principles of military necessity and proportionality.
88 Joint Chiefs of Staff, *Space Operations*, Joint Publication 3–14 (May 29, 2013), I–3.
89 Ibid., IV–3.
90 Corbett, *Some Principles of Maritime Strategy*, 128.
91 Raoul Castex, *Strategic Theories*, trans. and ed. Eugenia C. Kiesling (Annapolis, MD: Naval Institute Press, 1994), 235–237.
92 Bowen, *Spacepower and Space Warfare*, 194.
93 *Central mass* is the phrase used by both Clausewitz and Corbett.
94 Klein, *Space Warfare*, 91–92.
95 The choice of these terms is in keeping with Corbett's maritime strategy usage, although he does use the term *open* for *distant* on some occasions.
96 Joint Chiefs of Staff, *Space Operations* (May 29, 2013), II–8.
97 Ibid.
98 Carl von Clausewitz, "Die wichtigsten Grundsätze des Kriegführens zur Ergänzung meines Unterrichts bei Sr. Königlichen Hoheit dem Kronprinzen," in *Vom Kriege*, dritter Band (Berlin: Ferdinand Dümmler, 1832), 241.
99 Sun Tzu, *The Art of War*, 87.
100 Clausewitz, *On War*, 595.
101 Warden, "The Enemy as a System," 41–55.

102 Ibid., 43.
103 Ibid., 42.
104 Ibid., 44–49.
105 Wolfgang Wegener, *Naval Strategy of the World War*, trans. Holger H. Herwig (Berlin: E. S. Mittler & Sons, 1929; reprint, Annapolis, MD: U.S. Naval Institute Press, 1989).
106 Ibid., 30.
107 Ibid., 61.
108 Klein, *Space Warfare*, 80–87.
109 David E. Lupton, *On Space Warfare: A Space Power Doctrine* (Maxwell Air Force Base, AL: Air University Press, June 1988), 21.
110 Everett C. Dolman, *Astropolitik: Classical Geopolitics in the Space Age* (London: Frank Cass, 2002), 39, 71.
111 Bernard Brodie, *War and Politics* (New York: The Macmillan Company, 1973), 452.
112 Clausewitz, *On War*, 158.
113 Ibid., 141.
114 Corbett, *Some Principles of Maritime Strategy*, 5, 8.

Bibliography

Bacon, Francis. "Of the True Greatness of Kingdoms and Estates." In *The Essays*. 1601; reprint, Adelaide: The University of Adelaide, 2014.

Bowen, Bleddyn E. *Spacepower and Space Warfare: The Continuation of Terran Politics by Other Means*. PhD Thesis, Aberystwyth University, 2015.

Brodie, Bernard. *War and Politics*. New York: The Macmillan Company, 1973.

Brodie, Bernard. *Strategy in the Missile Age*. Santa Monica, CA: RAND Corporation, 1959.

Callwell, Charles E. *Military Operations and Maritime Preponderance: Their Relations and Interdependence*. With introduction by Colin S. Gray. London: William Blackwood and Sons, 1905; reprint, Annapolis, MD: Naval Institute Press, 1996.

Castex, Raoul. *Strategic Theories*. Translated and edited by Eugenia C. Kiesling. Annapolis, MD: Naval Institute Press, 1994.

Clausewitz, Carl von. *On War*. Translated and edited by Michael Howard and Peter Paret. Princeton, NJ: Princeton University Press, 1989.

Clausewitz, Carl von. *Principles of War*. Translated by Hans W. Gatzke. 1812; The Military Service Publishing Company, 1942; Clausewitz.com, 2016. www.clausewitz.com/readings/Principles/Clausewitz-PrinciplesOfWar-ClausewitzCom.pdf

Clausewitz, Carl von. *Vom Kriege*, erster Band. Berlin: Ferdinand Dümmler, 1832.

Clausewitz, Carl von. *Vom Kriege*, zweiter Band. Berlin: Ferdinand Dümmler, 1832.

Clausewitz, Carl von."Die wichtigsten Grundsätze des Kriegführens zur Ergänzung meines Unterrichts bei Sr. Königlichen Hoheit dem Kronprinzen." In *Vom Kriege*, dritter Band. Berlin: Ferdinand Dümmler, 1832.

Clodfelter, Mark A. "Molding Airpower Convictions: Development and Legacy of William Mitchell's Strategic Thought." In *The Paths of Heaven: The Evolution of Airpower Theory*, edited by Phillip S. Meilinger, 79–114. Maxwell Air Force Base, AL: Air University Press, 1997.

Corbett, Julian S. *Some Principles of Maritime Strategy*. London: Longmans, Green and Co., 1911; reprint, Annapolis, MD: Naval Institute Press, 1988.

Dolman, Everett C. *Astropolitik: Classical Geopolitics in the Space Age*. London: Frank Cass, 2002.

Douhet, Giulio. *The Command of the Air*. Translated by Dino Ferrari, 1921 and 1927; New York: Coward-McCann, 1942; reprint, Washington, DC: Air Force Museums and History Program, 1998.

European Commission. *Space Strategy for Europe*. Communication from the Commission to the European Parliament, the Council, the European Economic and Social Committee and the Committee of the Regions. October 26, 2016.

Eves, Stuart. "Angels and Demons: Cooperative and Non-cooperative Formation Flying with Small Satellites." Presentation. Surrey Satellite Technology Limited, University of Surrey, 2008.

Fadok, David S. "John Boyd and John Warden: Airpower's Quest for Strategic Paralysis." In *The Paths of Heaven: The Evolution of Airpower Theory*, edited by Phillip S. Meilinger, 357–398. Maxwell Air Force Base, AL: Air University Press, 1997.

Gray, Colin S. *Airpower for Strategic Effect*. Maxwell Air Force Base, AL: Air University Press, 2012.

Gray, Colin S. *Fighting Talk: Forty Maxims on War, Peace, and Strategy*. Westport, CT: Greenwood Publishing, 2007.

Gray, Colin S. Introduction to *Military Operations and Maritime Preponderance: Their Relations and Interdependence*. Charles E. Callwell. London: William Blackwood and Sons, 1905; reprint, Annapolis, MD: Naval Institute Press, 1996.

Gray, Colin S. *Weapons Don't Make War: Policy, Strategy, and Military Technology*. Lawrence, KS: University Press of Kansas, 1993.

Joint Chiefs of Staff. *Space Operations*. Joint Publication 3–14. April 10, 2018. www.jcs.mil/Portals/36/Documents/Doctrine/pubs/jp3_14.pdf.

Joint Chiefs of Staff. *Space Operations*. Joint Publication 3–14. May 29, 2013.

Klein, John J. *Space Warfare: Strategy, Principles and Policy*. Abingdon: Routledge, 2006.

Lianju, Jiang and Wang Liwen. *Textbook for the Study of Space Operations*. Beijing: Military Science Publishing House, 2013.

Lupton, David E. *On Space Warfare: A Space Power Doctrine*. Maxwell Air Force Base, AL: Air University Press, June 1988.

Mahan, Alfred T. *Sea Power in its Relation to the War of 1812: Volume I*. Boston, MA: Little & Brown, 1905.

Mahan, Alfred T. *The Problem of Asia and Its Effect Upon International Policies*. Boston: Little, Brown, and Company, 1900.

Mitchell, William. *Winged Defense: The Development and Possibilities of Modern Air Power — Economic and Military*. 1925; reprint, New York: Dover Publications, 1988.

Mitchell, William. *Our Air Force: The Keystone of National Defense*. New York: E.P. Dutton and Co, 1921.

Nayak, Michael. "Deterring Aggressive Space Actions with Cube Satellite Proximity Operations: A New Frontier in Defensive Space Control." *Air and Space Power Journal* vol. 31 no. 4 (Winter 2017): 92–102.

Nye, Jr., Joseph S. *Softpower: The Means to Success in World Politics*. New York: Public Affairs, 2005.

Office of the Assistant Secretary of Defense for Homeland Defense and Global Security. *Space Domain Mission Assurance: A Resilience Taxonomy*. September 2015. http://policy.defense.gov/Portals/11/Space%20Policy/ResilienceTaxonomyWhitePaperFinal.pdf?ver=2016-12-27-131828-623

People's Liberation Army Academy of Military Science, Military Strategy Studies Department. *Science of Military Strategy*. Beijing: Military Science Press, December 2013.

Pollpeter, Kevin L., Michael S. Chase, and Eric Heginbotham. *The Creation of the PLA Strategic Support Force and Its Implications for Chinese Military Space Operations.* Santa Monica, CA: RAND Corporation, 2017.

Sheldon, John B. and Colin S. Gray. "Theory Ascendant? Spacepower and the Challenge of Strategic Theory." In *Toward a Theory of Spacepower: Selected Essays*, edited by Charles D. Lutes and Peter L. Hays, with Vincent A. Mazo, Lisa M. Yambrick, and M. Elaine Bunn. Washington, DC: National Defense University Press, 2011.

Smith, M.V. "Spacepower and the Strategist." In *Strategy: Context and Adaption from Archidamus to Airpower*, edited by Richard J. Bailey Jr., James W. Forsyth Jr., and Mark O. Yeisley, 157–185. Annapolis, MD: Naval Institute Press, 2016.

Strassler, Robert B. *The Landmark Thucydides: A Comprehensive Guide to the Peloponnesian War.* New York: Free Press, 1996.

Sun Tzu. *The Art of War.* Translated by Samuel B. Griffith. Oxford: Oxford University Press, 1963.

Warden, John. "The Enemy as a System." *Airpower Journal* vol. 9 no. 1 (Spring 1995): 41–55. www.airuniversity.af.mil/Portals/10/ASPJ/journals/Volume-09_Issue-1-Se/1995_Vol.9_No1.pdf

Wegener, Wolfgang. *Naval Strategy of the World War.* Translated by Holger H. Herwig. Berlin: E.S. Mittler & Sons, 1929; reprint, Annapolis, MD: U.S. Naval Institute Press, 1989.

The White House. *The National Security Strategy of the United States of America.* December, 2017. www.whitehouse.gov/wp-content/uploads/2017/12/NSS-Final-12-18-2017-0905.pdf

Wylie, J.C. *Military Strategy: A General Theory of Power Control.* With introduction by John B. Hattendorf. New Brunswick, NJ: Rutgers University Press, 1967; reprint, Annapolis, MD: Naval Institute Press, 1989.

Ziarnick, Brent. *Developing National Power in Space: A Theoretical Model.* Jefferson, NC: McFarland, 2015.

3   Technology and space warfare

The security environment is also affected by *rapid technological advancements and the changing character of war.* The drive to develop new technologies is relentless, expanding to more actors with lower barriers of entry, and moving at accelerating speed.

James Mattis, 2018 U.S. National Defense Strategy[1]

Since the earliest times, humans have attempted to use technology to achieve decisive, strategic effects in war. Whether using the trireme, longbow, trébuchet, artillery, tank, or airplane, military strategists have modified operational style for the pursuit of victory in conflict. This condition will not change, as long as there is competition between states and groups in which violence and military action are viewed as viable solutions.

There is a danger in taking this approach to an extreme, however, where applying technology for military advantage or to achieve strategic effect takes primacy over all other considerations. Brian Hanley rightly notes, "A change in the tools of war is of minor consequence if a strategy is faulty or malignant or if commanders are deficient in moral and intellectual ability."[2] Consequently, it is imperative to remember that the application of technology should be guided by sound strategy.

For thousands of years, war and warfare have involved statesmen, strategy, violence, and the application of technology to aid in achieving political ends. Any future war initiated in or extending into space will involve the same considerations. In highlighting the enduring nature of war and technology's role, Baron Antoine-Henri de Jomini observed "… principles are unchanging, independent of the kind of weapons, of historical time and of place."[3] So, while technology will change and advance—thereby changing the character of war—war's fundamental nature remains the same. Consequently, historical experience and strategic writings that have stood the test of time can provide a useful framework for considering technology's place in war and warfare.

To make the implementation of a general space strategy a practical matter, an understanding of how strategy, tactics, and technology influence one another is needed by the strategist and military planner. Because tactics are closely tied to available technology, tactics may change with technology's advancement. Colin

Gray has observed, "Unless there are unusually powerful offsetting factors, it is generally true that weapons technology molds tactics and suggests operational style."[4] Even though the application of technology, or potential weapons systems, can achieve strategic effect, tactical actions and the application of technology—to include space technology—will most often impact the tactical level of war, and on occasion, the operational level as well. Furthermore, because technology affects the methods and means available in executing military operations, strategy may at times change as technology advances. Technology and its application in tactical action will help determine what strategies are considered practical for achieving success.

Although technological advantage is important when waging war, it is only one dimension in strategy and likely not the most important dimension.[5] Michael Handel has perhaps explained this best in describing technology's role in warfare:

> This point, like those that follow, is not intended to suggest that technology is unimportant. It is simply there to remind us that technology, while of the greatest importance, is still only the *means*; as such, it is always secondary to the political and strategic non-material dimensions of war. Thus, technology and material victories are inseparable from the political and 'strategic' dimensions, but in the final analysis they are at best only a necessary but rarely sufficient condition for a final and complete victory.[6]

People matter most

War is ultimately about people. Thucydides underscored this point by noting the interplay of people and polities' fear, honor, and interest in considering the decision to go to war.[7] While technology and technological advancement is important—especially for its potential in achieving strategic effect—people and their decisions for starting and ending wars rule supreme. J.C. Wylie wrote on the importance of people: "*The ultimate determinant in war is the man on the scene with the gun.* This man is the final power in war. He is control. He determines who wins."[8] Ralph Peters has similarly commented on the primacy of needing to understand people and their motivations by observing, "Technologies come and go, but the primitive endures.... In this age of technological miracles, our military needs to study mankind."[9] The enduring nature of war—to include why humans decide to go to war and what politically derived ends are to be achieved—underscores the need study and better understand basic human and societal interactions.

Moreover, when considering the importance of well-trained soldiers, marines, sailors, and airmen relative to the latest technological marvel, people again win out. Alfred Thayer Mahan argued this point in stating:

> Historically, good men with poor ships are better than poor men with good ships; over and over again the French Revolution taught this lesson, which

our own age, with its rage for the last new thing in material improvement, has largely dropped out of memory.[10]

Thus, it is not so much technology or the latest technological advancement that is important, rather what matters most is how people make use of such technology. Mao Tse-tung summarized this thinking succinctly: "Weapons are an important factor in war, but not the decisive factor; it is people, not things that are decisive."[11]

Technology and the legal regime

The current legal regime, specifically the 1967 Outer Space Treaty and the resulting assortment of multilateral agreements, has relevance when considering the application of technology in space, because the treaty serves as the legal precedent for the restriction of certain weapons and technologies.[12] Whether military actions are carried out commensurate with the understood meaning of the Treaty, which is sometimes not agreed upon by all, will be a decision for political and military leadership. One of the most widely referenced sections of the Treaty is Article IV, which declares,

> States Parties to the Treaty undertake not to place in orbit around the Earth any objects carrying nuclear weapons or any other kinds of weapons of mass destruction, install such weapons on celestial bodies, or station such weapons in outer space in any other manner.[13]

Because the treaty only mentions nuclear weapons and other kinds of weapons of mass destruction, other weapons are, in principle, not prohibited by the Outer Space Treaty.

Additionally, the Outer Space Treaty describes other restrictions that shape how weapons, technology, or military bases should be employed. Specifically, it describes:

> The Moon and other celestial bodies shall be used by all States Parties to the Treaty exclusively for peaceful purposes. The establishment of military bases, installations and fortifications, the testing of any type of weapons and the conduct of military manoeuvres on celestial bodies shall be forbidden. The use of military personnel for scientific research or for any other peaceful purposes shall not be prohibited. The use of any equipment or facility necessary for peaceful exploration of the Moon and other celestial bodies shall also not be prohibited.[14]

Of note, the Treaty does not define what is meant by *peaceful purposes*. It has been the consistent policy of the United States and other signatories that the term *peaceful purposes* is inclusive of defense and national security activities that are non-aggressive.[15] From the language above, military bases, installations, and fortification are permissible in orbit or in other locations, just not on celestial bodies.

Another United Nation agreement, the "Principles Relevant to the Use of Nuclear Power Sources in Outer Space," delineates considerations for the application of nuclear power technology in space. This agreement states:

> *Having considered* the report of the Committee on the Peaceful Uses of Outer Space on the work of its thirty-fifth session and the text of the Principles Relevant to the Use of Nuclear Power Sources in Outer Space as approved by the Committee and annexed to its report; *Recognizing* that for some missions in outer space nuclear power sources are particularly suited or even essential owing to their compactness, long life and other attributes; *Recognizing also* that the use of nuclear power sources in outer space should focus on those applications which take advantage of the particular properties of nuclear power sources....[16]

So, while nuclear weapons are prohibited per the Outer Space Treaty, nuclear power is considered well-suited for some space operations and missions.

Historical experience in applying new technologies

Although there is no guarantee of what the future entails, historical experience can teach valuable lessons regarding novel technologies and how they have influenced warfare. Such lessons can help guide the formulation of future space strategies. The historical examples discussed next include the maritime application of early aircraft, the use of submarines during both World Wars, nuclear deterrence during the Cold War, and philosophies on using technology for strategic advantage.

A word of caution is warranted before proceeding. Historical examples can be used to prove a multitude of things, many of which are conflicting in meaning. Consequently, it is unwise to attempt to establish a grand theory of innovation or create a model for explaining innovation.[17] Stephen Rosen has demonstrated the difficulties of such a task by analyzing much of the literature on military innovation, and he failed to find any patterns that would support using historical examples to explain innovation or postulate a grand theory.[18]

Despite this condition, strategists require answers and a practical understanding of historical experience. For this reason, there is often a persistent tension between historians and strategists because each may wish to use an understanding of past experiences differently. Nevertheless, within the context of technological innovation and how such technology has been used to achieve strategic effect, Colin Gray offers encouragement,

> Happily, it is not the task of the theorist to discourage the quest for improved military and strategic performance, quite the contrary, in fact. Rather [it] is the theorist's mission, at least with respect to maxims, to try to save people

of action both from themselves and from the seductive purveyors of the latest all but guaranteed way to win, and the like.[19]

Therefore, what will be presented here are maxims or truisms—concepts generally considered to be true—regarding technology's role in warfare, with the intent to re-teach what may have been forgotten by strategists, warfighters, and policy experts. Because it is possible to provide any number of historical examples to make a point regarding the outcomes of war, the examples that follow are not intended to be predictive of the future employment of technologies but are intended to elucidate how the development and employment of advanced technology can shape the strategic landscape and change the operational styles employed in warfare.

Maritime application of early aircraft

The early employment of aircraft is a revealing example of how the influence of technology had only slight initial effect on warfare. In the early twentieth century, aircraft were a relatively new technology, and how their employment could change the conduct of future wars was not well understood. One of the earliest writers of air power theory was Italian Air Marshal Giulio Douhet. In his book originally published in 1921, *The Command of the Air*, he contended that aircraft are the solution to strategic and tactical stalemates, and all future wars can be won from the air.[20] For Douhet, the aircraft's superior advantage is said to be its offensive characteristics—freedom of maneuver and speed—which are achieved by operating in the air.[21] Although Douhet recognized that land, sea, and air forces should cooperate to achieve common objectives, he placed special emphasis on each component achieving results independently.[22] As a consequence, air forces should operate and achieve results "to the complete exclusion of both army and navy."[23] Furthermore, he believed aircraft could achieve military victory without the efforts of the army or navy and, consequently, air forces are "first in order of importance" of all the armed services.[24]

Despite Douhet's prognostication and theory of air warfare, the early application of aircraft did not live up to his expectations. As is the case with many innovative technologies, any technological advancement is usually applied in a method consistent with the existing paradigm. Or, more simply, people generally attempt to employ new ideas or things in ways consistent with previous practice or operational style. As a result, the most significant advantages of new technology are not fully appreciated until sometime later.

This was indeed the case with the use of aircraft prior to the 1940s. According to the paradigm at the time, wars were won or lost by armies and navies; thus, aircraft should support those armies and navies. Based upon the writings of Alfred Thayer Mahan, navies achieve victory by seeking a decisive battle against the enemy's fleet, and the means of achieving this victory was through the battleship. Because the battleship was the centerpiece of offensive power at sea during the early twentieth century, it was the predominant view of U.S. naval officers of the time that the application of new aircraft technology should help the battle-ship perform its job better. This view was shared by the French naval officer, Raoul Castex. He wrote his book *Strategic Theories* from 1931–1939 and addressed the application of innovative technologies to naval strategy, including the aeroplane and the submarine. Castex showed great foresight by seeking to address the question whether airpower alone can achieve a decisive victory. He concluded that it cannot, at least not against naval forces.[25] To many naval strategists of that time, the early role of naval aircraft was envisioned to provide reconnaissance to expedite locating and engaging the enemy's fleet. Another early role for naval aircraft included providing airborne spotting and corrections for naval gunnery fire against other naval targets. Overall, the early uses of air-craft in the maritime domain were in modest, supporting roles.

Even during land warfare conducted by the U.S. Marine Corps, aircraft only played supporting roles before the 1940s. The U.S. Marine Corps' *Small Wars Manual* noted the utility of aircraft—because they operate from a position with tactical advantage—yet, aircraft were given missions supporting marines on the ground, including scouting, observing, and reconnaissance.[26] The manual states specifically, "The primary mission of combat aviation in a small war is the direct support of the ground forces."[27] Dedicated aviation attack was deemed necessary only when the enemy itself had a viable aviation threat or when bombings against enemy lines of communication and strongholds were needed.[28] Consequently, to marines of the time, there was nothing novel about aircraft when it came to fighting small wars and, as a result, aircraft were used in a manner commensurate with a traditional strategy supporting expeditionary land forces. Despite the technological advancement that the aircraft represented, the strategy of the U.S. Marine Corps failed to take full advantage of the aircraft's capability, as recognized in more modern times.

Submarine warfare during the two World Wars

Although submarine warfare is a subset of maritime warfare, the early German successes in using submarines during both World Wars show how superior tech-nology, if not countered, may be exploited for military advantage. Indeed, Germany viewed the submarine as a major determinant for victory, as has been noted, "The intended instrument of decision in war for Germany in 1917 and again in late 1942 and early 1943 was the U-boat."[29] German military leadership viewed the employment of submarines as directly affecting the outcome of the war. Even though Germany enjoyed early successes in both wars using U-boats, history has repeatedly shown that superior technology and tactics can eventually be countered, at least to some extent, and the submarine was no exception.

World War I

During 1917–1918, Germany employed the concept of *guerre de course*—or commerce raiding—using unrestricted submarine warfare. As an island nation,

Great Britain relied on maritime commerce coming into her ports, and German military leadership exploited this fact by using single, patrolling U-boats to sink ships headed towards the British Isles. The strategy was so successful that the situation looked very bleak for the British by April 1917, and it appeared as though 40 U-boats along the western approaches to the British Isles would starve the British into submission.[30] In February and March of 1916, 1,149 ships entered British ports, while during the same period the following year, the number was less than 300.[31] To many observers of the day, it looked as though Germany would win unless British shipping losses could be stopped.

One of the reasons German U-boats enjoyed resounding success in attacking merchant shipping along British sea lines of communication was that their U-boats operated counter to the naval warfare theories of the day, including the sea power ideas of Mahan. The conventional wisdom within both British and U.S. navies said that battleships reigned supreme, and victory at sea was achieved by first defeating the enemy's battleships and then defeating lesser cruisers that attacked maritime trade. Under Mahanian thinking, this order of battle would protect the merchant shipping and ensure the flow of seaborne commerce. However, Germany's naval leaders found a way around this doctrinal approach. Submarines were not cruisers but were instead raiders that could not easily be found. Thus, the German U-boats' success at sea did not depend on battleships or the battle-fleet. The German military leaders' ability to modify the "rules of the game" and operational style through their unconventional use of submarines changed the strategic landscape and sent the British leadership reeling to find a way to counter the U-boat threat.

To protect shipments arriving in Britain, especially coming through the western approaches, a counter-strategy of the convoy was employed. The convoy counter-strategy meant that groups of merchant cargo ships travelled together, being protected against U-boat attack by naval escort vessels. The object of the escorts was not to sink enemy U-boats but to protect the convoy during its transit. As a result, convoying was a defensive strategy and was not seeking a decisive battle. Escorts had only to defend and control that part of the ocean over which the convoy was travelling, and only for the length of time that they were in transit.

By the end of 1917, it was apparent that Germany's strategy of unrestricted submarine warfare had failed, and with this failure Germany had lost its chance for a quick victory. Germany maintained their U-boat strategy, however, and extended the campaign into U.S. waters in May 1918. Despite adding bigger and more capable U-boats, Germany never found a way to overcome the convoy defense. The Germans expended more effort and cost, while sinking fewer ships.

World War II

During World War II, Germany adjusted their operational style by employing U-boats for unrestricted submarine warfare through the coordinated actions of a group or "wolf pack." This adjustment was in response to the success of the convoying countermeasure. Admiral Karl Dönitz of the German navy developed the concept of the wolf pack, while also employing new tactics such as short-range torpedo attacks, to improve the chance of a kill when engaging a convoy. Dönitz believed in focusing on the potential gross tonnage when attacking shipping, instead of a ship's immediate military value, and he thought the war could be decided by submarines alone.[32] The sinking of shipping's gross tonnage capacity was all that mattered under his strategy. No distinction was made between the number of ships and qualitative value, and therefore large, empty ships would be preferred over smaller ships carrying cargo.[33] This new wolf pack tactic employed six to 12 U-boats operating together to target large merchant ships, whether empty or cargo laden, and the U-boats would operate in locations where German intelligence determined convoys would transit.

Because Dönitz believed enemy shipping should be seen as a collective whole, wolf packs marauded off the coast of the U.S. eastern seaboard after it became more difficult to sink Allied shipping in the North Atlantic. The wolf packs initially had remarkable success operating there, and the United States lost sea control directly off its own shore for a time. During the first half of 1942, over 2.3 million tons of shipping capacity was sunk in the western hemisphere, with most of that being lost along the eastern and Gulf coasts of the United States.[34] The U.S. Navy countered the wolf pack by again employing convoys and using escorts with offensive capabilities to mitigate the U-boat threat off the eastern seaboard of the United States.[35]

The German wolf packs initially enjoyed stunning successes, and their use might have been decisive in the early 1940s. In the end, however, Dönitz's strategy of sinking tonnage failed, in part due to Allied convoying and increases in shipbuilding. The U.S. Ship-For-Victory program reached a production goal of three Liberty Ships per day, and each ship could be built in less than two months.[36] Beginning in July 1942, the United States built more ships than the Germans sank.[37] Besides the ship building effort, the capability to decipher German coded messages and radar-equipped Allied patrol aircraft also directly contributed to countering the U-boat threat.[38]

Overall, the history of the German U-boat experience in both World Wars demonstrates that technological advancement and exploitation can lead to a tactical and operational advantage, which can contribute in deciding a war's outcome. Nevertheless, historical experience also shows that, with any technological advancement, a counter to such advancement will likely follow. Such is the natural progression of technological advancement and warfare.

Nuclear weapons during the Cold War

With the development and use of nuclear weapons against Japan in 1945 to bring World War II to an end, America had unprecedented military capability and economic strength, which resulted in the United States establishing an unrivalled international influence. However, four years after Hiroshima and Nagasaki, the Soviet Union followed suit and developed its own nuclear weapons. Leaders in both the United States and the Soviet Union believed that nuclear weapons were directly linked to a country's national power. Even today, leaders from countries such as North Korea and Iran perceive nuclear technology as providing countries with greater diplomatic influence. History suggests that other countries have pursued nuclear technology for similar reasons.

In highlighting their importance at the strategic-level war, Thomas Mahnken notes, "Indeed, the nuclear revolution represents the most clear-cut case of technology affecting the conduct of war in recent centuries."[39] The development of nuclear weapons is noteworthy owing to its stabilizing influence between the United States and the Soviet Union. Because of the mutual devastation that would result in a nuclear exchange, both superpowers avoided direct military confrontation and instead participated in "proxy" conflicts in Korea, Vietnam, and Afghanistan. During the 1962 Cuban Missile Crisis, the possibility of a full-scale nuclear exchange contributed to Khrushchev's decision to remove Soviet missiles from Cuba.

Many lessons can and have been drawn from the Cold War. For the purposes of space strategy, a key lesson from this period suggests that nuclear technology—or any weapons technology considered as being the most devastating—may at times deter or dissuade others considering direct military action.[40] At times, such deterrence may be stabilizing among competing states, lending to peace among the international community. A question worth asking, of which there is disparate views among security experts, is whether the actual capability demonstration of such devastating weapons is required to have a deterrent effect on par with nuclear weapons. This question will likely be relevant when considering future military systems or potential weapons in space.

Revolution in military affairs, transformation, and Third Offset

Within the U.S. national security community in recent decades, there is a consistent theme of wanting to use advanced technology to affect the waging of war. Thomas Mahnken has noted,

> Reliance on advanced technology has been a central pillar of the American way of war, at least since World War II. No nation in recent history has placed greater emphasis upon the role of technology in planning and waging war than the United States.[41]

In 1946, Walter Lippman wrote on this idea—within the context of the new atomic bomb and the rockets to deliver them. To Lippman, the arrival of atom bombs and rockets appeared as:

> the perfect fulfillment of all wishful thinking on military matters: here is war that requires no national effort, no draft, no training, no discipline, but only money and engineering know-how of which we have plenty. Here is the panacea which enables us to be the greatest military power on earth without investing time, energy, sweat, blood, and tears....[42]

Examples from more recent decades of seeking to use technology for military advantage include the ideas of the *Military Technical Revolution* and the resulting *Revolution in Military Affairs*, which gained popularity in the early 1990s. Some ardent supporters argued that the information revolution marked a complete break with the past. One 1993 report predicted, "The Military Technical Revolution has the potential fundamentally to reshape the nature of warfare. Basic principles of strategy since the time of Machiavelli ... may lose their relevance in the face of emerging technologies and doctrines."[43]

While technological advances are prerequisite for these revolutions, technology alone is insufficient. Critical of the idea of technology changing the nature of war, Andy Marshall—head of the DoD Office of Net Assessment at the time—thought that true revolutions take place only when the armed forces develop new concepts of operation and create new organizations. In his view, the key task facing the armed forces at the time was not to rush out and purchase new equipment, but to figure out the most appropriate conceptual innovations and organizational changes.[44] Similar to Marshall's observations, James Fitz-Simonds and Jan van Tol note that history suggests three common preconditions for realizing a full Revolution in Military Affairs: technological development; doctrinal innovation; and organization adaptation.[45]

Transformation

A follow-on to the Revolution in Military Affairs concept was Transformation. *Transformation* refers to a revolutionary or significant improvement in hardware, tactics, or doctrine, and this term gained popularity in the early 2000s.[46] Proponents of military transformation believed militaries that evolve gradually are susceptible to being overtaken by adversaries willing to risk all on revolutionary changes.[47] It is thought that the concept of transformation gained a large following within the U.S. defense establishment due, in part, to the fear of the unknown.[48]

With the fall of the Soviet Union and the end of the decades-long Cold War, U.S. military planners did not know where the next threat would come from. Without knowing one's future enemy, it would be extremely difficult to formulate multi-year defense appropriation budgets and advocate expensive weapons systems programs. According to the prevailing logic of transformation, U.S. supremacy over a near-peer competitor or future threat could be virtually ensured by embracing technology as the solution for the defense establishment's inability to determine the next threat. As a result, a long-term plan can be made for defense appropriations, and a strategy of transformation can be enacted. Unfortunately, to what end appropriations are to be made and against whom transformational technologies are to be employed, very few could discern. Transformation advocates routinely sought a military force that was lighter, more mobile, and easily deployable to emerging crisis spots around the world.

The Third Offset

The idea of the *Third Offset* was formally announced by the U.S. Secretary of Defense Chuck Hagel in late 2014. In an official memorandum, he advocated for a "third offset strategy that puts the competitive advantage firmly in the hands of American power projection over the coming decades."[49] The term *offset* referred to a strategy seeking to use novel technology or approaches to achieve a competitive advantage. As described by some defense department officials, there have been two previous "offsets" in U.S. history since the end of World War II. The first U.S. strategic offset is said to have occurred in the early 1950s at the start of the Cold War, and it sought to blunt Soviet numerical and geographical advantage along the inner-German border by introducing, demonstrating, and developing the operational and organizational constructs for employing nuclear weapons on the battlefield.[50] The second U.S. strategic offset was said to have occurred in the 1970s and 1980s through coordinated and networked precision strike, stealth, and surveillance capability for conventional forces to affect negatively Soviet military effectiveness.[51]

In Hagel's original memo, this Third Offset approach sought defense department innovation in several, linked areas:

- integration of leadership development practices with emerging opportunities to rethink how the defense department develops both managers and leaders;
- a new long-range research and development planning program to identify, develop, and field breakthrough technologies and systems that sustain and advance the U.S. military power;
- a reinvigorated war gaming effort to develop and test alternative ways of achieving U.S. strategic objectives and help the Defense Department think more clearly about the future security environment;
- new operational concepts to explore how to employ resources to greater strategic effect and deal with emerging threats in more innovative ways; and
- continued examination of business practices and finding ways to be more efficient and effective through external benchmarking and focused internal reviews.[52]

In all, Hagel's approach to accelerated innovation included both the human, organizational, and technological elements to seek a strategic advantage over potential future adversaries. Later in 2015, U.S. Deputy Secretary of Defense Bob Work further described the Third Offset activities to include six broad areas: anti-access and area-denial, guided munitions, undersea warfare, cyber and electronic warfare, human-machine teaming, and war gaming and development of new operating concepts.[53]

Critics of the Third Offset Strategy have said the concept lacks clarity and simplicity, resulting in criticisms that it tries to be everything, which causes it to be nothing of practical value.[54] In recognizing the criticisms of the concept, Bob Work explained, "So what we want to do is develop successive generations of many warfighting capabilities. The technology is never, never the definitive answer. You have to be able to incorporate those technologies into new operational and organizational constructs."[55] He further advised that the United States will need to demonstrate these innovative capabilities to convey that any attempt by a potential adversary to achieve operational success is likely to fail, even if an adversary achieves an initial advantage in time and space.[56]

These past initiatives within the U.S. defense community of revolution in military affairs, transformation, and the Third Offset sought to achieve a military advantage through the novel application of technology. Such thinking is commensurate with the commonplace insight that from time to time there is a radical change in the character and conduct of warfare.[57] While historical experience demonstrates that novel methods of employing technology may change the operational style or achieve strategic effect when employed, policy-makers and strategists are cautioned not to place an over-reliant faith in the wonders of advanced technology.

Technology's use in space warfare

Based upon historical experience, technology should be expected to play a significant role in the conduct of space warfare, especially at the tactical level and potentially in shaping operational style. Many of the ideas that follow have been demonstrated during the conduct of military operations or are part of a well established, general understanding of war and warfare. Even though a large or unlimited war in space has not occurred—thankfully—certain ideas for space warfare's conduct can be suggested. The observations on technology's role in space strategy that follow will be relevant to all space powers—whether they be great, medium, or emerging powers. When considering space warfare and technology's role, Michael Handel's observation is foundational: "Technology is only a means in war, which cannot produce complete victory and success by itself."[58]

Some space power proponents may advocate making space-based attack the centerpiece of any military strategy, just as sea and air power advocates have in the past. Because space-based weapons may seem beyond the reach of attack by a potential adversary and the effects from space-based weapons may be devastating, it is expected that "space power" will be considered an attractive doctrinal approach for considering military operations. The most strident of space power advocates could argue that space warfare and advanced space-based technology have fundamentally nullified the historical theory and principles of warfare, as described by Thucydides, Sun Tzu, and Clausewitz. Space warfare and advanced space-based weapons, however, will not change the nature of warfare, but will affect the character of how war is conducted. At the heart of warfare are people and their competitive struggle, and many—if not all—of the time-honored lessons of war remain relevant, even with significant advances in technology.[59]

The best advice this strategist has for those within the defense community or military services seeking to use the latest innovative technology to guarantee victory or define a new way of war is "Please, stop it." Within the U.S. defense community especially, there is a history of the periodic re-emergence of technology's application as a "hot" and "new" strategic concept. Colin Gray warns of the succession of purportedly strategic concepts *du jour* within the U.S. defense community that have repeatedly gained popularity, and then official endorsements, based on a largely false promise of superior performance.[60] He warns that "there will always be a market for new sounding ideas expressed in jargon and neatly acronymic. They come, they go, and they reappear in slightly different guise in the future."[61] Gray advises current and future strategists that there are just three defenses against the usually false—at least exaggerated—strategic promise of the hot, new concept: common sense, experience, and a sound education in strategy, especially in the timeless works of Thucydides, Sun Tzu, and Clausewitz.[62]

Space-based assets can be stabilizing

Advanced space-based technology, including associated weapons, can have a stabilizing effect on the international community. As was the case with nuclear weapons during the Cold War, if a weapons system poses a large enough threat to two or more adversaries, its potential use can cause state leaders to avoid direct confrontation and escalation in hostilities. This point is not to suggest that future space-based weapons will eliminate tensions among competing states, nations, or groups, but is rather to highlight that technologically advanced weapons can provide a stabilizing influence at times.

The statement of U.S. President Lyndon Johnson in 1967 exemplifies this point of advanced capability promoting stability, even though Johnson's context is on reconnaissance capability vice nuclear technology. He told a group:

> We've spent $35 or $40 billion on the space program. And if nothing else had come out of it except the knowledge that we gained from space photography, it would be worth ten times what the whole program has cost. Because tonight we know how many missiles the enemy has and, it turned out, our guesses were way off. We were doing things we didn't need to do. We were building things we didn't need to build. We were harboring fears we didn't need to harbor.[63]

Friction and uncertainty are not eliminated

> What I fear is not the enemy's strategy, but our own mistakes.
>
> Thucydides[64]

Some advocates have claimed that advanced technology and its employment during modern warfare obviate the need for those defensive strategies meant to handle "friction" and "uncertainty." When discussing the rapid advances in information and information-related technologies, the U.S. National Defense

Panel of 1997 stated that technological advantages could "dissipate the fog of war."[65] While technology may offer opportunities to reduce the fog of war, viewpoints that suggest it can be eliminated are simply incorrect. Despite technological advances, fog and friction remain fundamental to the nature of warfare.

Historical experience has shown that ambiguity, miscalculation, incompetence, and chance are all ingredients during times of war. It is expected that these factors will play a role in warfare in space as well. Barry Watts has noted that technology will not solve the problem of chance, uncertainty, and miscalculation in observing: "Human limitations, informational uncertainties, and nonlinearity are not pesky difficulties better technology and engineering can eliminate, but *built-in or structural* features of the violent interactions between opposing polities pursuing incommensurable ends we call war."[66] It is not expected that space warfare employing the latest technologies will be any different in this regard. Despite the many advantages of employing the latest innovation, space-based or space-enabled technology will not eliminate friction, chance and uncertainty, but may at times only reduce it. Technophiles believing that superior technology will enable one to know everything that is happening in all areas of interest will be sadly disappointed. Even if it were possible to monitor and collect intelligence on every aspect possible, such a condition would not guarantee knowing an enemy's actual intentions.

Space technology will not win wars alone

Technology has its place in deciding the outcome of any battle, which in turn has a bearing on the outcome of wars and conflicts in general. An application of superior firepower may be employed to obliterate and demoralize enemy forces, thereby achieving strategically decisive results. Although technology and its destructive application during tactical operations can achieve such an outcome, usually many tactical actions occur during an operational campaign, and an entire war may be composed of several campaigns. Therefore, one decisive victory through the application of superior technology does not mean a war is won.

During World War II, Winston Churchill examined Royal Air Force war plans and disagreed with the view that the aerial bombing of Germany was a guaranteed path to victory. Churchill told his advisers that he deeply mistrusted "these cut and dried calculations, which showed infallibly how the war would be won."[67]

For the development of space strategy, this reality means that the application of space-based technologies is unlikely to win wars by itself. While space-related technology may provide a tactical advantage in combat and influence operations style, it is unlikely that a single technology will decide the outcome of a conflict. Technology should not be used in isolation from the overall strategic objectives of the war, and "technological proficiency is no substitute for strategic acuity."[68] While using space-based technology as part of a broad military strategy improves the likelihood of achieving lasting results, any war plan that relies on one specific application of space-based or space-reliant technology to achieve victory is an unbalanced and ill-conceived approach to strategy.

Historical experience illustrates that despite technological advancements in firepower or capability, any superior space-based weaponry will eventually be countered—or at least mitigated—by adversaries. Commenting on the natural progression in warfare where technological capability begets a countermeasure, Clausewitz observes, "If the offense were to invent some major new expedient … the defensive would also have to change its methods."[69] Although space-based weapons or the effect of space-enabled terrestrial forces may be devastating, an adversary will develop or reverse-engineer a comparable weapon, steal similar capabilities, or even find a counter to a weapon's effectiveness through new tactics, techniques, and procedures. Advancement, countermeasure, and counter-countermeasure: that is the natural cycle of technology in warfare.

Space warfare will not be simple or easy

Because warfare involves conflict between determined and impassioned belligerents, any promise of a quick path to victory through employing a single advanced technology or means of attack should be met with skepticism. Winston Churchill warned, "Never, never, never believe any war will be smooth and easy, or that anyone who embarks on the strange voyage can measure the tides and hurricanes he will encounter."[70] This same lesson holds for space warfare.

Some space technology advocates could argue that space-based attack will be simple or easy because few or no military personnel may be directly affected during conflicts in space. Only satellites, anti-satellite weapons, or space-based weapons would be targeted for attack, so there is little downside risk to conducting military operations in space. Also, the frequently used adage, "Satellites don't have mothers," implies that war in space will have fewer drawbacks when compared to terrestrial conflict because there is no loss of human life in space. Space technology proponents may describe how simple a space-based attack would be against an unprepared or unsuspecting enemy. Thomas Mahnken observers the downside of such thinking in writing,

> Washington's penchant for advanced technology also fostered the illusion among some that the United States could use force without killing American soldiers and innocent civilians, and among America's enemies the impression that the United States was averse to sustaining casualties.[71]

Ultimately, any promises of a simple and easy path to victory in space will be mostly empty and far from reality because most enemies will adapt and react when attacked.

To those who overestimate their own capability and underestimate the enemy's, Mao's warning rings true:

> In the end, Mr. Reality will come and pour a bucket of cold water over these chatterers, showing them up as mere windbags, who want to get things on the cheap, to have gains without pains.… There is no magic short-cut.[72]

Conclusion

The influence of technology on the conduct of warfare and the development of strategy is still not fully understood within many military communities. This misunderstanding may indeed be true of technology's role in the development and execution of space strategy. Based upon historical experience, it can be expected that advances in space-related technology will be used initially in ways commensurate with the current military and operational paradigms. Therefore in the near-term, space operations will likely continue to play mostly supporting roles—albeit important ones—to operations on land, at sea, in the air, and in cyberspace. It may be some time until the strategic advantages of space-based or space-enabled operations are fully understood and effectively employed. While advances in space-related technology or space-based weaponry will not change the fundamental nature of warfare, these advances will change warfare's conduct and character.

Technology is important in war. There is a proper balance and perspective that must be sought, and Thomas Mahnken advises on the need to strike the right balance:

> If the enthusiasts are guilty of hyping technology, the skeptics have all too often discounted the role of technology in war. Although technology is not the only—or necessarily the most important—determinant of success, its effects should not be ignored.[73]

Mahnken notes that evolutionary advancements in precision guidance and stealth technologies are two examples where applying advanced technology had far-reaching strategic consequences.[74] A balanced understanding of technology's influence on the conduct of military operations can lead to the development of a more complete general theory of space strategy and suggest future operational style.

It is expected that space operations 200 years from now will look significantly different from space operations of today. To get a hint of how space operations of the future will be different from today, one can compare maritime operations during the Age of Sail to maritime operations of today. Just over two centuries ago, transoceanic shipping travelled using primarily the seasonal prevailing winds, and shipping that tried to deviate from the prescribed seasonal trade routes was at risk of taking an excessive amount of time to reach an intended destination, or not reaching the destination at all. It was not until the use of coal-fueled steam engines that transoceanic shipping was at last permitted to transit without being restricted by seasonal wind patterns.

Just as oceanic travel of the past was dictated by seasonal wind patterns, many space operations of today are determined primarily by orbital mechanics, or the gravitational pull of celestial bodies. In the future, when propulsion technology advances to the point where extended space travel is possible using more efficient sources of abundant energy—such as fusion reactors or advanced electric propulsion drives—it is expected that space travel will increase exponentially. Furthermore, improved propulsion technology will allow a state's interests in space to move beyond just near-Earth concerns and extend to cislunar regions and beyond.

Notes

1 Department of Defense, *2018 National Defense Strategy of the United States of America: Sharpening the American Military's Competitive Edge* (2018), 3, emphasis original, www.defense.gov/Portals/1/Documents/pubs/2018-National-Defense-Strategy-Summary.pdf
2 Brian Hanley, "Transformation Ballyhoo," *U.S. Naval Institute Proceedings* vol. 132.9.1 (September 2006), 67.
3 Baron Antoine-Henri de Jomini quoted in Crane Brinton, Gordon A. Craig, and Felix Gilbert, "Jomini," in *Makers of Modern Strategy: Military Thought from Machiavelli to Hitler*, ed. Edward Meade Earle (Princeton, NJ: Princeton University Press, 1948), 84. Originally cited in *Traité des grandes operations militaires*, volume 3 (Paris: 1804–1806), 333. Similar thoughts are in Antoine-Henri de Jomini, *The Art of War* (1862; reprint, London: Greenhill Books, 1992), 17 and 347.
4 Colin S. Gray, *Weapons Don't Make War: Policy, Strategy, and Military Technology* (Lawrence, KS: University Press of Kansas, 1993), 78.
5 David J. Lonsdale, *The Nature of War in the Information Age: Clausewitzian Future* (London: Frank Cass, 2004), 53.
6 Michael I. Handel, *Masters of War: Classical Strategic Thought*, 3rd ed. (London: Frank Cass, 2001), xxi. Emphasis original.
7 Robert B. Strassler, *The Landmark Thucydides: A Comprehensive Guide to the Peloponnesian War* (New York: Free Press, 1996), 43.
8 J.C. Wylie, *Military Strategy: A General Theory of Power Control*, with introduction by John B. Hattendorf (New Brunswick, NJ: Rutgers University Press, 1967; reprint, Annapolis, MD: Naval Institute Press, 1989), 72. Emphasis original.
9 Ralph Peters, *Fighting for the Future: Will America Triumph?* (Mechanicsburg, PA: Stackpole Books, 1999), 171–172.
10 Alfred Thayer Mahan, *Influence of Sea Power Upon the French Revolution and Empire, 1793–1812* (Boston: Little, Brown, and Co., 1892), 102; as referenced in Colin S. Gray, *Fighting Talk: Forty Maxims on War, Peace, and Strategy* (Westport, CT: Greenwood Publishing, 2007), 96.
11 Mao Tse-tung, *Selected Military Writings of Mao Tse-tung* (Peking: Foreign Language Press, 1963), 217–218.
12 United Nations General Assembly, resolution 2222 (XXI), *Treaty on Principles Governing the Activities of States in the Exploration and Use of Outer Space, including the Moon and Other Celestial Bodies*, or *The Outer Space Treaty* (1967), www.unoosa.org/oosa/en/ourwork/spacelaw/treaties/outerspacetreaty.html
13 *Outer Space Treaty*, Article IV.
14 Ibid., Paragraph 2, Article IV.
15 The White House, *National Space Policy of the United States of America* (June 28, 2010), 3, www.nasa.gov/sites/default/files/national_space_policy_6–28–10.pdf
16 United Nations General Assembly, resolution 47/68, *Principles Relevant to the Use of Nuclear Power Sources in Outer Space* (December 14, 1992), Preamble, www.unoosa.org/oosa/en/ourwork/spacelaw/principles/nps-principles.html
17 Williamson Murray and Allan Millett, introduction to *Military Innovation in the Interwar Period* (Cambridge: Cambridge University Press, 1996), 4.
18 Stephen Peter Rosen, "Thinking about Military Innovation," in *Winning the Next War: Innovation and the Modern Military* (Ithaca, NY: Cornell University Press, 1991), 1–53.

19 Gray, *Fighting Talk*, 42.
20 Giulio Douhet, *The Command of the Air*, trans. Dino Ferrari (1921 and 1927; New York: Coward-McCann, 1942; reprint, Washington, DC: Air Force Museums and History Program, 1998), 15–29.
21 Ibid., 15.
22 Ibid., 4.
23 Ibid., 5.
24 Ibid., 29.
25 Raoul Castex, *Strategic Theories*, trans. and ed. Eugenia C. Kiesling (Annapolis, MD: Naval Institute Press, 1994), 321.
26 United States Marine Corps, *Small Wars Manual* (Washington, DC: 1940; reprint, 1990), section 9–1.
27 Ibid., section 9–23.
28 Ibid., section 9–25.
29 Colin S. Gray and Roger W. Barnett, *Seapower and Strategy* (Annapolis, MD: Naval Institute Press, 1989), 11.
30 George W. Baer, *One Hundred Years of Sea Power: The U.S. Navy 1890–1990* (Stanford CA: Stanford University Press), 67.
31 Ibid.
32 Ibid.
33 Ibid., 192.
34 Ibid., 194.
35 Marc Milner, "Anglo-American Naval Cooperation in the Second World War, 1939–45," in *Maritime Strategy and the Balance of Power: Britain and America in the Twentieth Century*, eds. John B. Hattendorf and Robert S. Jordan (New York: St. Martin's Press, 1989), 250.
36 Baer, *One Hundred Years of Sea Power*, 199–200.
37 Ibid., 201.
38 Ibid., 193, 198 and 201.
39 Thomas G. Mahnken, *Technology and the American War of War Since 1945* (New York: Columbia University Press, 2008), 223.
40 Colin S. Gray, *Modern Strategy* (Oxford: Oxford University Press, 1999), 308–309.
41 Mahnken, *Technology and the American Way of War Since 1945*, 5.
42 Walter Lippmann, "Why Are We Disarming Ourselves?" *Redbrook Magazine* (September 1946), 106. As referenced in Lawrence Freedman, *The Evolution of Nuclear Strategy*, 3rd ed. (New York: Palgrave Macmillan, 2003), 45.
43 Michael J. Mazarr, Jeffrey Shaffer, and Benjamin Ederington, "The Military Technical Revolution: A Structural Framework" (The Center for Strategic and International Studies, 1993), 28.
44 Mahnken, *Technology and the American Way of War Since 1945*, 5.
45 James R. FitzSimonds and Jan M. van Tol, "Revolutions in Military Affairs," *Joint Force Quarterly* (May 1994), 25–26. http://ndupress.ndu.edu/portals/68/Documents/jfq/jfq-4.pdf
46 Department of Defense, *U.S. Quadrennial Defense Review* (September 30, 2001), IV, http://archive.defense.gov/pubs/qdr2001.pdf
47 Greg Jaffe, " 'New and Improved?' Special Report: Spending for Defense," *The Wall Street Journal*, March 28, 2002.
48 It has been surmised that much of the enthusiasm for military transformation stems from the "yearning for military certainty." *See* Colin S. Gray, *Defining and Achieving Decisive Victory* (U.S. Army War College: Strategic Studies Institute, April 2002), 24.
49 Chuck Hagel, Department of Defense, *Memorandum on the Defense Innovation Initiative* (November 15, 2014), http://archive.defense.gov/pubs/OSD013411–14.pdf

50 Bob Work, "Center for New American Security Defense Forum" (JW Marriott, Washington, DC, December 14, 2015), www.defense.gov/News/Speeches/Speech-View/Article/634214/cnas-defense-forum
51 Ibid.
52 Hagel, *Memorandum on the Defense Innovation Initiative*, 1–2.
53 Work, "Center for New American Security Defense Forum."
54 Andy Massie, "Reframing the Third Offset as a 21st-Century Model for Deterrence," *War on the Rocks*, March 28, 2016, https://warontherocks.com/2016/03/reframing-the-third-offset-as-a-21st-century-model-for-deterrence/
55 Bob Work, "Remarks by Deputy Secretary Work on Third Offset Strategy" (Brussels, Belgium, April 28, 2016), www.defense.gov/News/Speeches/Speech-View/Article/753482/remarks-by-d%20eputy-secretary-work-on-third-offset-strategy/
56 Work, "Remarks by Deputy Secretary Work on Third Offset Strategy."
57 Gray, *Fighting Talk*, 64.
58 Handel, *Masters of War*, xxiv
59 "Regardless of what character a war assumes, it is always a human activity." See Lonsdale, *The Nature of War in the Information Age*, 36.
60 Gray, *Fighting Talk*, 65.
61 Ibid.
62 Ibid.
63 Quoted in William E. Burrows, *Deep Black: Space Espionage and National Security* (New York: Random House, 1986), vii.
64 Thucydides, *History of the Peloponnesian War* (432 BC), 1.144.1
65 National Defense Panel, *Transforming Defense: National Security in the 21st Century* (Washington, DC: U.S. Government Printing Office, 1997), iv, as quoted in Mahnken, *Technology and the American Way of War Since 1945*, 178. Additionally, William Owens stated "This revolution challenges the hoary dictums about the fog and friction of war." William Owens with Ed Offley, *Lifting the Fog of War* (New York: Farrar, Straus, and Giroux, 2000), 15.
66 Barry W. Watts, "Clausewitzian Friction and Future War," McNair Paper No. 68 (National Defense University, 2004), 78. Emphasis original.
67 Winston Churchill quoted in Elliot A. Cohen, "Churchill and Coalition Strategy in World War II," in *Grand Strategies in War and Peace*, ed. Paul Kennedy (New Haven, CT: Yale University Press, 1992), 66. Original citation comes from a Defence Committee (Operations) meeting of January 13, 1941.
68 Mahnken, *Technology and the American Way of War Since 1945*, 6.
69 Carl von Clausewitz, *On War*, trans. and eds. Michael Howard and Peter Paret (Princeton, NJ: Princeton University Press, 1989), 362.
70 Winston Churchill, quoted in Cohen, "Churchill and Coalition Strategy in World War II," 66.
71 Mahnken, *Technology and the American Way of War since 1945*, 6.
72 Mao Tse-tung, *Selected Military Writings*, 218–219.
73 Mahnken, *Technology and the American Way of War since 1945*, 220.
74 Ibid., 227.

Bibliography

Baer, George W. *One Hundred Years of Sea Power: The U.S. Navy 1890–1990*. Stanford CA: Stanford University Press.
Brinton, Crane, Gordon A. Craig, and Felix Gilbert. "Jomini." In *Makers of Modern Strategy: Military Thought from Machiavelli to Hitler*, edited by Edward Meade Earle, 77–92. Princeton, NJ: Princeton University Press, 1948.

Burrows, William E. *Deep Black: Space Espionage and National Security.* New York: Random House, 1986.

Castex, Raoul. *Strategic Theories.* Translated and edited by Eugenia C. Kiesling. Annapolis, MD: Naval Institute Press, 1994.

Clausewitz, Carl von. *On War.* Translated and edited by Michael Howard and Peter Paret. Princeton, NJ: Princeton University Press, 1989.

Cohen, Elliot A. "Churchill and Coalition Strategy in World War II." In *Grand Strategies in War and Peace,* edited by Paul Kennedy, 43–70. New Haven, CT: Yale University Press, 1992.

Department of Defense. *2018 National Defense Strategy of the United States of America: Sharpening the American Military's Competitive Edge.* 2018.

Department of Defense. *U.S. Quadrennial Defense Review.* September 30, 2001. http://archive.defense.gov/pubs/qdr2001.pdf

Douhet, Giulio. *The Command of the Air.* Translated by Dino Ferrari. 1921 and 1927; New York: Coward-McCann, 1942; reprint, Washington, DC: Air Force Museums and History Program, 1998.

FitzSimonds, James R. and Jan M. van Tol. "Revolutions in Military Affairs." *Joint Force Quarterly* (May 1994): 24–31. http://ndupress.ndu.edu/portals/68/Documents/jfq/jfq-4.pdf

Freedman, Lawrence. *The Evolution of Nuclear Strategy.* 3rd edition. New York: Palgrave Macmillan, 2003.

Gray, Colin S. *Fighting Talk: Forty Maxims on War, Peace, and Strategy.* Westport, CT: Greenwood Publishing, 2007.

Gray, Colin S. *Defining and Achieving Decisive Victory.* U.S. Army War College: Strategic Studies Institute, April 2002.

Gray, Colin S. *Modern Strategy.* Oxford: Oxford University Press, 1999.

Gray, Colin S. and Roger W. Barnett. *Seapower and Strategy.* Annapolis, MD: Naval Institute Press, 1989.

Gray, Colin S. *Weapons Don't Make War: Policy, Strategy, and Military Technology.* Lawrence, KS: University Press of Kansas, 1993.

Hagel, Chuck. Department of Defense. *Memorandum on the Defense Innovation Initiative.* November 15, 2014. http://archive.defense.gov/pubs/OSD013411-14.pdf

Handel, Michael I. *Masters of War: Classical Strategic Thought.* 3rd edition. London: Frank Cass, 2001.

Hanley, Brian. "Transformation Ballyhoo." *U.S. Naval Institute Proceedings* vol. 132.9.1 (September 2006): 64–68.

Jaffe, Greg. "'New and Improved?' Special Report: Spending for Defense." *The Wall Street Journal.* March 28, 2002.

Jomini, Antoine-Henri de. *The Art of War.* 1862; reprint, London: Greenhill Books, 1992.

Jomini, Antoine-Henri de. *Traité des grandes operations militaires,* volume 3. Paris: 1804–1806.

Lippmann, Walter. "Why Are We Disarming Ourselves?" *Redbrook Magazine* (September 1946).

Lonsdale, David J. *The Nature of War in the Information Age: Clausewitzian Future.* London: Frank Cass, 2004.

Mahan, Alfred Thayer. *Influence of Sea Power Upon the French Revolution and Empire, 1793–1812.* Boston: Little, Brown, and Co., 1892.

Mahnken, Thomas G. *Technology and the American War of War Since 1945.* New York: Columbia University Press, 2008.

Mao Tse-tung. *Selected Military Writings of Mao Tse-tung.* Peking: Foreign Language Press, 1963.

Massie, Andy. "Reframing the Third Offset as a 21st-Century Model for Deterrence." *War on the Rocks.* March 28, 2016. https://warontherocks.com/2016/03/reframing-the-third-offset-as-a-21st-century-model-for-deterrence/

Mazarr, Michael J., Jeffrey Shaffer, and Benjamin Ederington. "The Military Technical Revolution: A Structural Framework." The Center for Strategic and International Studies, 1993.

Milner, Marc. "Anglo-American Naval Cooperation in the Second World War, 1939–45." In *Maritime Strategy and the Balance of Power: Britain and America in the Twentieth Century,* edited by John B. Hattendorf and Robert S. Jordan, 243–270. New York: St. Martin's Press, 1989.

Murray, Williamson and Allan Millett. Introduction to *Military Innovation in the Interwar Period, 1–5.* Edited by Williamson Murray and Allan Millett. Cambridge: Cambridge University Press, 1996.

National Defense Panel. *Transforming Defense: National Security in the 21st Century.* Washington, DC: U.S. Government Printing Office, 1997.

Owens, William, with Ed Offley. *Lifting the Fog of War.* New York: Farrar, Straus, and Giroux, 2000.

Peters, Ralph. *Fighting for the Future: Will America Triumph?* Mechanicsburg, PA: Stackpole Books, 1999.

Rosen, Stephen Peter. *Winning the Next War: Innovation and the Modern Military.* Ithaca, NY: Cornell University Press, 1991.

Strassler, Robert B. *The Landmark Thucydides: A Comprehensive Guide to the Peloponnesian War.* New York: Free Press, 1996.

The White House. *National Space Policy of the United States of America.* June 28, 2010. www.nasa.gov/sites/default/files/national_space_policy_6-28-10.pdf

Thucydides. *History of the Peloponnesian War.* 432 BC.

United Nations General Assembly. Resolution 2222 (XXI). *Treaty on Principles Governing the Activities of States in the Exploration and Use of Outer Space, including the Moon and Other Celestial Bodies,* or *The Outer Space Treaty.* 1967. www.unoosa.org/oosa/en/ourwork/spacelaw/treaties/outerspacetreaty.html

United Nations General Assembly. Resolution 47/68. *Principles Relevant to the Use of Nuclear Power Sources in Outer Space.* December 14, 1992. www.unoosa.org/oosa/en/ourwork/spacelaw/principles/nps-principles.html

United States Marine Corps. *Small Wars Manual.* Washington, DC: 1940; reprint, 1990.

Watts, Barry D. "Clausewitzian Friction and Future War." McNair Paper No. 68. National Defense University, 2004.

Work, Bob. "Center for New American Security Defense Forum." JW Marriott, Washington, DC, December 14, 2015. www.defense.gov/News/Speeches/Speech-View/Article/634214/cnas-defense-forum

Work, Bob. "Remarks by Deputy Secretary Work on Third Offset Strategy." Brussels, Belgium, April 28, 2016. www.defense.gov/News/Speeches/Speech-View/Article/753482/remarks-by-d%20eputy-secretary-work-on-third-offset-strategy/

Wylie, J.C. *Military Strategy: A General Theory of Power Control.* With introduction by John B. Hattendorf. New Brunswick, NJ: Rutgers University Press, 1967; reprint, Annapolis, MD: Naval Institute Press, 1989.

4   Space deterrence and the law
of war

War has an enduring nature. Therefore, millennia of historical experience and
the practical implementation of strategy can help highlight the relationship
between deterrence and the Law of War in space. Through such foundational
understanding, more suitable space strategies may be developed, and effectual
technological solutions proposed to achieve political ends during conflict.

This chapter will address the broad family of thinking that includes the ideas
of *assurance, compellence, deterrence*, and *dissuasion. These ideas pertain to
affecting the decision calculus of others*, including the desire to affect (not necessarily change) the thought processes of potential friends and adversaries. While
it is sometimes easy for policy-makers and strategists to argue about the definitions of terms associated with this idea, it is postulated here that if the idea is
considered as merely seeking ways to affect another's thinking, the concept and
methods to achieve the desired end state are easier to consider and develop holistically. This is because when using specific definitional language, it is easy to
develop unintentional "gaps and seams." Furthermore, while words have
meaning, there is much "intellectual baggage" associated with the previous
terms, to the point that it is, at times, difficult to have a thoughtful and objective
conversation on how these concepts relate to space strategy.

Space deterrence

When it is desired to affect others' thinking to avoid direct confrontation and for
them to believe that hostilities should not be pursued because of expected failure
or associated costs, this is commensurate with *deterrence* (through either denial
or punishment). This may entail affecting—to include changing or reinforcing—
the decision calculus of the potential adversary. In a frequently cited definition
by Thomas Schelling, *deterrence* is persuading a potential enemy that it is in his
own interests to avoid certain courses of activity.[1] To Schelling, deterrence is
like the defense, or passive, because it is based on a response to something considered unacceptable.[2] The purpose of deterrence is to influence someone's
behavior.[3]

The underlying basis of space deterrence theory—a subset of general
deterrence—is that the threat of credible and potentially overwhelming force or

other retaliatory action against any would-be adversary is sufficient to deter most potential aggressors from conducting hostile actions in space. This definition may also be referred to as *deterrence by punishment*. In contrast, when the idea is to convey to an adversary to cease some current action—requiring the adversary to respond—this is more the role of *compellence*.[4] Schelling described *compellence* as a direct action that persuades an opponent to give up something that is desired.[5] Any effort to affect the decision calculus of another is best served by clearly communicating one's desire, intent, capability, and rational for military response.[6] This requisite communication is not achieved solely through official statements or policy documents, but also through a demonstrated history of consistent actions.

Of note, both military and non-military means are applicable in affecting the thinking of others. These non-military means equate to the soft power, or the diplomatic, informational, and economic instruments of national power. Non-military means can be used to affect another state leader's thought processes—whether reinforcing a currently held view that is beneficial to the affecting state or changing the view of another state's leadership or polities. Consequently, a practical implementation may entail political and diplomatic efforts, such as new international treaties or agreements; multimedia stories presenting news in a favorable perspective; and commerce and trade activities that increase one's own economic influence or negatively affect a potential adversary or opposing alliance.

James Finch and Shawn Steene have noted the need to think about *space deterrence* as deterring attacks against space systems while bolstering an overarching deterrence posture.[7] They suggest an approach utilizing the familiar means of imposing cost, denying benefit, and encouraging restraint. Through such an approach, it is thought that should deterrence fail in space, national leaders have options and capabilities that allow them to prevail in the broader terrestrial conflict.[8]

Some critics may question whether there is, in fact, *space deterrence* or if the idea should just be called *deterrence*, implying that there is only one multi-domain war to be deterred.[9] While intending to be thoughtful, this question misses the point. A better question to ask is whether current activities and systems in space can change the thought processes of potential adversaries. This answer is simple, at least to this strategist: "Yes." Words having meaning and any terminology should be as clear as possible. Yet any phrase or terminology chosen to convey the concept is of secondary importance to an understanding that there are indeed actions that can be taken relative to space that affect the decisions of others. Moreover, there are actions relative to the instruments of national power and operations in the other domains that can affect decisions relative to operations and actions in space.

Comparisons to nuclear deterrence

Because thinking of space as a warfighting domain is a relatively new idea, some policy-makers and strategists have sought to pull from other frameworks to think about great power competition in space. As a result, there are frequent comparisons of space deterrence to nuclear deterrence during the Cold War. Admittedly, there is a range of disparate views of what nuclear deterrence actually means. In one such view on the purpose of nuclear weapons, Bernard Brodie writes, "Thus far the chief purpose of our military establishment has been to win wars. From now on its chief purpose must be to avert them. It can have almost no other useful purpose."[10] According to Brodie, nuclear weapons only exist to prevent wars, not to be used during them.

Finch and Steene have addressed comparing nuclear and space deterrence, while being careful to note the ways in which the two are different. In describing the role of nuclear deterrence, they write:

> Deterrence had existed previously, of course, but the unprecedented destructive power of atomic weapons made the price of deterrence failure unaffordable. Scholars, particularly in the United States, spent careers studying and theorizing about various aspects of the superpowers' military balance—first-strike stability, escalation ladders, and conditions for deterrence failure. By the end of the Cold War, the United States had generally accepted a theory of deterrence that sought to ensure strategic stability by assuring, in the event of deterrence failure, the total annihilation of the opponent.[11]

The authors note the problem with using the nuclear model for thinking about space. Unlike nuclear weapons—which could threaten the extinction of mankind—space weapons are viewed as any another weapon rather than as weapons that represent the pinnacle of conflict or that define bilateral relationships. While Finch and Steene observe that there is no effective defense against a large-scale nuclear attack, this is not true for space. Consequently, the efficacy of deterrence in a space context may vary based on both weapon and target, creating a situation where deterrence holds for some targets while simultaneously failing for others.

In countering the applicability of nuclear deterrence to space deterrence, Karl Mueller argues that nuclear deterrence and space deterrence are not parallel concepts, despite having similarities.[12] He says the unique operating environment and physics of orbital mechanics create an operational and strategic environment in which conventional wisdom does not apply, going as far as to suggest space deterrence may not be a useful construct at all. Likewise, James Lewis asserts that concepts of deterrence developed from nuclear weapons are not applicable to space assets, because nuclear weapons are uniquely destructive and that the bipolar global conflict was a unique political moment in international affairs.[13]

Michael Krepon has also written on this issue of seeking comparisons and has defined *space deterrence* as "deterring harmful actions by whatever means against national assets in space and assets that support space operations."[14] In contrast, he defines *nuclear deterrence* as "deterring harmful actions by means of nuclear weapons."[15] Because the concept of space deterrence is not well developed, Krepon suggests using the better understood concept of nuclear deterrence to help better inform the idea of deterrence in space.[16] He does note that the concept of nuclear deterrence never reached a consensus view on the requirements for deterrence to be effective, to include force structure, and he concludes, as have others, that the answer to how much nuclear capability is enough for deterrence is "it depends."[17]

In comparing space deterrence to nuclear deterrence, Krepon suggests that some of the same initiatives that proved successful in Cold War nuclear deterrence may prove useful in space deterrence. In noting the potentially common areas of overlap between space deterrence and nuclear deterrence, he says:

> The key elements of space deterrence, as with nuclear deterrence, are secure retaliatory capabilities sufficient to deny advantages to an attacker, effective command and control mechanisms, and redundant safety and security mechanisms to prevent accidental as well as unauthorized use of military capabilities. In addition, successful deterrence requires situational awareness, attribution capabilities, as well as resilient space assets so that the United States is able to identify the perpetrator of harmful actions and continue to utilize space for national and economic security despite these acts.[18]

Because of the broad-scope taken when considering the functions needed, Krepon's list of common areas is useful for considering those capabilities needed for space deterrence to be effective.

Assurance, alliances, and extended deterrence

The objective to affect the decision calculus of others includes reinforcing another's thinking in ways considered beneficial to oneself. For instance, if a country is considered a good ally and partner, there may be a desire to reinforce this partnership and demonstrate the continued need to not proliferate weapons of mass destruction, while being part of an extended deterrence umbrella arrangement. Such an approach would be part of a strategy incorporating *assurance* between allies as part of a collective or bilateral security agreement. Within the context of nuclear weapons, the U.S. Quadrennial Defense Review from 2001 included the idea of assurance, which stated, "America's alliances and security relations give assurance to U.S. allies and friends and pause to U.S. foes. These relationships create a community of nations committed to common purposes."[19] Consequently, efforts seeking to reinforce cooperative relationships that include promoting common interests are thought to be beneficial in deterring aggression by potential adversaries.

Assurance has been part of the U.S. extended deterrence approach with the North Atlantic Treaty Organization (NATO), along with bilateral agreements with Japan and South Korea. Key for considering the effectiveness of U.S. extended deterrence is understanding that its effectiveness depends on how both allies and potential adversaries perceive the credibility of U.S. commitments. Moreover, the perceptions of allies and potential adversaries will not be uniform and can vary extensively based upon historical, cultural, and other unique circumstances.[20]

In noting the complexities of assurance and influencing the perception of others regarding extended deterrence, Colin Gray has noted:

> To extend deterrence it is not sufficient simply to have the capability to reach the putative enemy. That enemy must believe that he runs an unacceptably large risk of suffering intolerable pain should the extended deterrent ever be unleashed against him. Credibility alone does not ensure a sufficiency of deterrent or strategic effect. Indeed, there is a fundamental tension between credibility and prospective pain. Because of sensible fear of retaliation, the more painful an action is, the less likely it is to be taken, and the less likely it is that anyone will believe it will be taken.[21]

For this reason, there needs to be a belief that the political will exists to respond with severe military response if attacked. For the United States and European countries, a credible response may be shaped by the Law of Armed Conflict.

Because the idea of deterrence appears to have applicability in the space domain, concepts such as *extended deterrence* may prove useful in the future as part of mutual defense treaties or bilateral agreements. Consequently, extended space deterrence may afford assurance for allies and partners through the protection of common space-related interests, to include the potential use of force in support of collective self-defense arrangements. Dean Cheng aptly observes, however, that all countries do not view extended deterrence the same. Cheng notes that China will likely seek to employ all its various forces and capabilities in a holistic manner in pursuit of its ends, meaning that deterrence solely within the space domain is not typically a consideration. As a result, space powers, like the United States, should understand that some countries may view extended deterrence as embodying all its national capabilities, including land, sea, air, outer space, cyber, and nuclear forces.[22]

Alliances can influence access to and use of space. Therefore, alliances are important for space security and in attempting to achieve some level of space deterrence. In noting the advantages for Australia working with the United States on common space objectives through an alliance, Steve Henry of the Royal Australian Air Force has noted, "The main thing I have learned is that working with allies is challenging, but it is very much worth the effort. History will show that alliances built on shared values, mutual respect and complimentary strength win."[23] In Henry's view, no other domain requires the same degree of global access and cooperation that space does; consequently, a mix of international partners is needed to achieve common objectives in space. Allies and partners bring a myriad of benefits with respect to space operations, with one of the most beneficial being diverse geography. For example, the United States has space situational awareness sharing agreements with at least 17 countries, and having space-observation systems spread across the globe directly improves space situational awareness efforts.[24] Also, allies may have established relationships with various countries within a region, which others may find useful. For example, Henry notes that Australia is in the Indo-Pacific region, and its proximity to the established powers in this region, especially China and India, may provide insights that are valuable to others, like the United States.[25]

In describing the benefits coming from alliances, Gregory Schulte contends that, at a strategic level, the North Atlantic Treaty Organization (NATO) is uniquely positioned to bolster deterrence in space. This is because the alliance is increasingly reliant on space for its collective defense and economic prosperity, and an attack on the space assets of any one ally impacts the security of all allies.[26] Schulte observes that NATO is dependent on space, while asserting that its doctrine and planning have not kept up. He says NATO should continue to build the expertise and capacity to conduct operations enabled by space; ensure that doctrine, requirements, and planning account for the operational advantages provided by space; and adapt exercises and training to ensure forces can effectively exploit space-based capabilities.[27]

The strategic benefits coming from alliances have long been noted, to include the writings of Thucydides and Sun Tzu. Because of the inherent advantages from having allies and forming coalitions, some may seek to embrace a strategy designed to divide such alliances or coalitions.

The law of war and the inherent right of self-defense

> Nothing ... shall impair the inherent right of individual or collective self-defense if an armed attack occurs....
>
> Article 51 of the United Nations Charter[28]

Within many Western defense communities' perspectives, deterrence is frequently considered most effective if there is a credible threat of retaliatory action or force in response to a hostile act. Yet, establishing credibility is not an easy task. In recognition of the need to have a credible threat of military consequences, Thomas Schelling has noted, "Saying so, unfortunately, does not make it true; and if it is true, saying so does not always make it believed."[29] What is considered a credible action following an armed attack is typically governed by the Law of Armed Conflict (LOAC), which is sometimes also referred to as the Law of War. While not intended to be directive of any future action, the ideas and principles within the LOAC have relevance when considering responses to a hostile act and armed attack in space. Therefore, within the American or Western style of war, the LOAC has significance in shaping what is considered a reasonable and justifiable response, consequently the LOAC affects deterrence theory.

The LOAC has been defined as "that part of international law that regulates the conduct of armed hostilities."[30] It is based on two main sources: the first is customary international law arising out of hostilities and binding for all states; and the second is international treaty law, which impacts only those states having ratified a particular agreement. The inherent right of self-defense serves as the foundation of the Law of Armed Conflict. This right applies during peace or war and stems from customary international law dating back at least 300 years. Furthermore, this right is delineated in Article 51 of the United Nations Charter, of which an excerpt is provided at the beginning of this section. The purpose of the LOAC is to reduce the damage and casualties of any conflict; protect combatants and noncombatants from unnecessary suffering; safeguard the fundamental rights of combatants and noncombatants; and make it easier to restore peace after the conflict's conclusion.

The LOAC addresses many of the issues regarding the reasons to go to war and what is considered appropriate use of force. Even though self-defense can be used to justify military action under customary international law and international treaty law, collective self-defense can also be used for justification as well. Under Article 51 of the United Nations Charter, collective self-defense may be invoked. This means that if a state which is part of a cooperative defense agreement is attacked, then those other states being part of the same cooperative agreement can act against a belligerent, even though they themselves were not attacked.[31] Such collective defense agreements have been used between states for centuries and have contributed to international stability.

When considering space strategy, two principles contained in the LOAC are most significant: these are the principles of lawful targeting and military necessity.[32] Together, these principles help form the basis to consider damage resulting from an attack on a legitimate military objective before the act occurs.

The principle of lawful targeting, which is inclusive of the principle of distinction, is based on three underpinnings.[33] First, a belligerent's right to injure the enemy is not unlimited. Second, launching attacks against civilian populations is prohibited. Third, distinctions between combatants and noncombatants must be made to spare injury to noncombatants as much as possible. Consequently, under lawful targeting, all "reasonable precautions" must be taken to ensure only military objectives are targeted, so that damage to civilian objects (collateral damage) or death and injury to civilians (incidental injury) are avoided as much as possible.[34] Military objectives are combatants and those objects which, by their nature, location, purpose, or use, effectively contribute to the enemy's war-fighting or war-sustaining capability. Additionally, civilians and civilian objects may not be made the object of attack. Civilian objects consist of all civilian property and activities, other than those supporting or sustaining the enemy's war-fighting capability.

The second principle—military necessity—calls for using only that degree and kind of force required for the partial or complete submission of the enemy, while considering the minimum expenditure of time, life, and physical resources.[35] This principle is designed to limit the application of force to that required for carrying out lawful military purposes, and is also referred to as the principle of proportionality. Sometimes, this principle is misunderstood and misapplied to support the excessive and unlawful application of military force, because military necessity could be incorrectly argued to justify the accomplishment of any mission. Although the principle of military necessity recognizes that some collateral damage and incidental injury to civilians may occur when a legitimate military target is attacked, this does not excuse the wanton destruction of lives and property disproportionate to the military advantage to be gained.

Rules of engagement

Regarding the desire to affect the decision calculus of another through the threat of credible action, the next area of consideration is Rules of Engagement, particularly those as defined by the U.S. military. While the Rules of Engagement are not considered strictly law or a legal basis, the rules are shaped by the LOAC. Therefore, the Rules of Engagement seek to implement military action under a legal regime considered acceptable under customary international law and international treaty law. These rules help shape the understanding of what is considered an appropriate use of force during either peace or conflict by a state's or coalition's fighting forces.

It should be expected that war in space will observe many of the same restrictions as warfare in the other domains of war. Within the Unites States, the "rules" are subdivided into two sub-categories: Standing Rules of Engagement and Supplemental Rules of Engagement. Standing Rules of Engagement provide overarching guidance for the application of force during peace and war.[36] The Chairman of the U.S. Joint Chiefs of Staff promulgates the U.S. Standing Rules of Engagement, describing three types of self-defense. First, national self-defense applies to the United States, its forces and, in specific circumstances, its nationals and their property. Second, collective self-defense applies to designated non-U.S. forces, foreign nationals, and their property. Third, unit self-defense applies to a particular military element, including individuals and other forces in the vicinity.[37]

In contrast, the U.S. Supplemental Rules of Engagement are issued for the accomplishment of mission objectives during specified military actions or operations. The Supplemental Rules of Engagement typically delineate what is considered mission essential equipment, which may apply to equipment or property considered vital for the accomplishment of the mission. What makes these supplement rules noteworthy is that mission essential equipment—a physical asset—may be deemed necessary to protect by force because of its importance. This interpretation would seem to be especially relevant to the space domain, where highly valuable and mission essential satellites are unmanned, and loss of life may not be a direct concern.

Implications for space strategy

The connection between the LOAC and space strategy has four aspects. First, the inherent right of self-defense applies to satellites and other critical space systems. Some policy-makers may question if this right applies to satellites, under the adage that "satellites don't have mothers."[38] If no human life is directly threatened by an armed attack in space, then, in theory, there is no need to protect or defend satellites through military means. The counter to this question is found under Article 2(4) of the United Nations Charter, which describes the need to refrain from the threat or use of force against a state's territorial integrity—which may be interpreted as including a state's physical property.[39] Additionally, under the U.S. Chairman of the Joint Chiefs of Staff Standing Rules of Engagement, national self-defense and collective self-defense are defined as applying to both persons and property.[40] Recent public comments by U.S. military leadership, including U.S. Strategic Command Commander General John Hyten, also support the view that the right of self-defense applies in space.[41]

Second, considering the principles of targeting and necessity, it is not unlawful to cause collateral damage to the natural environment during an attack upon a legitimate military objective. There is an obligation, however, to avoid unnecessary damage to the environment—to the extent that it is practicable to do so—consistent with mission accomplishment. Destruction of the natural environment not necessitated by mission accomplishment and carried out wantonly is prohibited, and the environmental damage resulting from an attack on a legitimate military objective should be considered ahead of time. For space operations, this means creating orbital debris to achieve military objectives is thus permissible. That said, during war, means should be employed to protect and preserve the natural environment in space, and the destruction of the orbital environment through debris generation not necessitated by mission accomplishment and carried out recklessly is prohibited. For these reasons, any anticipated orbital debris resulting from an attack on a legitimate military objective should be considered during targeting analysis and selection.

Third, during future conflicts, there may be the need to not only target a specific satellite but a specific subsystem on that satellite. Under the LOAC, targeting distinction between multiple hosted payloads or subsystems on a single satellite is likely needed. Because today's commercial satellites may have multiple paying customers with different hosted payloads on each satellite, the principle of lawful targeting conveys the need to target only a specific subsystem on a satellite relating directly to the military objective. For example, the military may use commercial satellites for some communications, and so only that subsystem being the military object should be considered during the targeting process. This same idea of specific targeting also holds when considering jamming and interference of signal bandwidth used for military purposes. Admittedly, current technology and capabilities make targeting and engaging only a single sub-system on a satellite difficult. Additionally, conducting battle damage assessment after such an attack may likely prove challenging.

Fourth, adherence to the principle of military necessity does not preclude responding to an armed attack in space in a different domain. Consequently, if a system or asset considered essential is attacked in space, a response can include actions on land, at sea, or in the air. Needing to protect national interests in space through military action means potentially risking human life to defend and protect critical space assets. Hence, those seeking to protect national interests in space may need to put service members in harm's way, and a legitimate response to a hostile act under the LOAC may cause loss of life. Certainly, responding to a hostile action through a different domain could be perceived as escalatory by an adversary; therefore, appropriate messaging regarding intent, objectives, and capabilities should be conducted well before the potential onset of hostilities.

Preemption or anticipatory self-defense

There are few strategic concepts as hotly debated as anticipatory self-defense—or preemption. This is particularly the case when considering military action in space. Colin Gray observes, "Preemption is not controversial; legally, morally, or strategically."[42] Gray states this because preemption is based upon hundreds of years of customary international law and treaty law. Despite this historic precedence, most space powers have much to do in developing technical capabilities and communicating defense policies before preemption in space is, in fact, a practical and viable means of protecting national interests. These discussions are especially needed as competition in space grows. While being perhaps counterintuitive, developing the concepts of preemption well before conflict occurs enhances deterrence and may promote international peace and stability.

Preemption is an offshoot of the inherent right of self-defense, which pertains to a state being able to defend itself in response to an armed attack. As noted previously, Article 51 of the Charter of the United Nations recognizes nothing should impair the applicability of a state's inherent right of self-defense.[43] In contrast, *anticipatory self-defense* occurs before an armed attack or hostile act has actually occurred.

Admittedly, the legitimacy of anticipatory self-defense is debated among legal scholars. Some legal experts take a restrictive interpretation of Article 51 of the U.N. Charter by stating that the language "armed attack occurs" connotes self-defense only after an attack has begun or happened. Anticipatory or preemptive action would, accordingly, be illegitimate under the Charter.

In contrast, the other legal scholars take a different view, stating the U.N. Charter's language does not impair the inherent right of anticipatory self-defense under customary international law and is thereby permissible under certain conditions. Unlike treaties, customary international law is not created by what states put down in writing but, rather, by what states do in practice. Secretary of State Daniel Webster's case writings in 1842 regarding the *Caroline* diplomatic crisis and Roberto Ago's legal writings in 1980 similarly conclude preemption is a legitimate action under the conditions of necessity, proportionality, and immediacy.[44] Those conditions mean that there must be a need for military action against an adversary, the response should be proportional given the threat, and that the threat is considered to be of an immediate nature. The decision to act preemptively is a political choice, supported with military capability and justified to the international community.

One of the best-known—and still considered controversial by many security professionals—security documents advocating the legitimacy of preemption is the 2002 U.S. National Security Strategy. It is notable in its unabashed consideration of preemptive action, as well as a preventative war, when there is an imminent threat to America. The document states, "The United States has long maintained the option of preemptive actions to counter a sufficient threat to our national security."[45] The strategy articulates that the greater the threat, the greater the risk of inaction, thereby justifying anticipatory action to defend U.S. interests, even if uncertainty remains as to the time and place of the enemy's attack. Therefore, to prevent hostile acts by adversaries, the United States will—if necessary—act preemptively.

Preemption in space will need to meet the three preconditions—necessity, proportionality, and immediacy—as noted previously. However, while preemption in space should not be considered controversial, space as a domain of warfare presents special challenges for preemptive action. Specifically, preemption in space necessitates capabilities and processes for "observing" what is occurring, "categorizing" potential threats, and "communicating" understandings with the international community.[46]

Observing

The goal of preemption is to act before an adversary, to mitigate or minimize a threat before negative consequences are realized. Acting inside an adversary's decision cycle is commensurate with concepts developed by John Boyd, specifically acting within an adversary's orient, observe, decide and act (OODA) loop. Boyd held the belief that an ability to perform the OODA loop faster than an adversary was the key to victory.[47] Similarly, preemptive action is also aimed at acting faster than an adversary to achieve one's objectives.

Yet, to act preemptively requires the ability to observe the actions of others in space, or have extensive space situational awareness. A comprehensive, real-time situational awareness capability is required to orient and act within the space environment. Without such situational awareness, it seems doubtful that anticipatory self-defense can be executed optimally. Many countries, including China, Russia, and the United States, have determined the need to improve awareness of operations in the space domain and are investing in broad space situational awareness capabilities. For instance, the Commander of Air Force Space Command, General John Raymond, has said regarding U.S. space capabilities, "We have four geosynchronous space situational awareness program satellites that actually drift just below the GEO belt, kind of the neighborhood watch for space."[48] Raymond describes these satellites as augmenting the ground-based radar that the U.S. Air Force uses to track objects in orbit, giving a comprehensive picture of what is going on beyond the atmosphere.[49]

Categorizing

The decision to act preemptively is ultimately a political decision, informed by the magnitude and manner in which forces or assets are being threatened.

Without such an informed decision, preemption is merely aggression. With the growing number of space capabilities—to include those that could be considered as threatening—the decision to protect space capabilities is considered as increasingly urgent. The requisite information helps determine whether the military necessity precondition is met, a concept that requires a general under-standing of what constitutes an "armed attack" (U.N. Charter Article 51 lan-guage), "threat or use of force" (U.N. Charter Article 2(4) language), or "hostile act or demonstrated hostile intent" (U.S. Joint Chiefs of Staff Standing Rules of Engagement language). While the previous terminologies have similarities, each has a slightly different context and meaning. At times, these differences have resulted in ambiguity in when self-defense, let alone preemption, in space warfare is considered consistent with either customary international law or treaty law.

Another challenge for preemption in space is a result of the broad range of possible actions. This includes activities ranging from reversible (jamming) to non-reversible (destroying critical satellite electronics), and includes kinetic or non-kinetic actions, such as an anti-satellite missile or a laser. This range of potential offensive actions to critical space systems has contributed to confusion on what types of imminent armed attack or hostile acts rise to the occasion to warrant preemption. Some strategists may question whether non-kinetic and reversible actions necessitate a military response. For these reasons, clarity is needed in military doctrine, as well as any Rules of Engagement, regarding the conditions justifying self-defense and informing a decision to be made by polit-ical leadership for preemptive action in space.

Communicating

The final consideration in weighing preemptive action is communicating. A country's policy regarding preemption, with its emphasis on the legal basis for preemption and how armed attack and hostile acts in space are considered, needs to be shared among the international community. Such dialogue with allies and partners will lessen any misunderstanding or uncertainty during a war extending into space. By communicating what types of situations will meet necessity, pro-portionality, and immediacy—considerations for preemptive action—a space power will be better able to garner international support for potential coalitions or alliances to address emerging threats. Moreover, communicating a preemptive policy potentially makes it seem a more credible recourse—as viewed by poten-tial adversaries—with an acknowledgment of the challenges of convincing others or causing others to believe necessarily what is said.

Cautions against preemption

Some security experts may assess that preemptive action is not worth the associ-ated cost, and therefore, should be eliminated from the strategy toolkit. For example, Elbridge Colby states:

[A] space defense strategy that relied excessively, let alone exclusively, on striking an adversary's counterspace assets preemptively could thus put the nation in an impossible political-military position, one in which it would be required to strike early in a crisis to ensure it could attack a potential adversary's counterspace architecture before they dispersed or readied their defenses. It seems clear that no American political leader would want to be forced into such a position, and with ample reason.[50]

Brian Chow, however, refutes Colby's view, saying that Colby's preemption is not anticipatory self-defense at all, but more akin to a first-strike against an "adversary's counterspace architecture before they have dispersed or readied their defenses."[51] Chow sees Colby's perspective as being drawn upon nuclear deterrence theory and the use of a first nuclear strike that can significantly, if not totally, disable an opponent's second nuclear strike capability, which is destabilizing and dangerous.[52]

When considering the pros and cons of preemption, the strategist should remember the following: preemption is not a decision whether to go to war or not; it is about the terms under which the conflict will occur. The decision for war has already been made.[53] When a country preempts, it makes a choice between either receiving the first blow or striking first. In most cases, deciding against preemption when certain hostile action is imminent is unlikely to improve one's military position, because the situation dictates that the decision for war has already been made. If the attack is certain, there are only two reasons for withholding the use of preemptive force. First, it may not be feasible to preempt because one's OODA loop is not adequate, including the inability to accurately observe or identify the existence of the imminent threat. Without such an informed decision, preemption is merely aggression. Second, it may be judged politically important to allow the enemy to attack first, thereby branding himself unquestionably as the aggressor.[54]

Dissuasion

Another aspect of a holistic space strategy seeking to influence the decision calculus of potential adversaries is *dissuasion*, which is meant to discourage the initiation of military competition.[55] Often, the term *dissuasion* is used when describing actions "that should be taken against those identified as posing a threat to American interests prior to such potential adversaries having the actual capability to pose a danger."[56] To be effective, dissuasion activities must occur before a threat manifests itself. Dissuasion includes "shaping activities," which are typically nonmilitary in scope and are conducted during peacetime. Within the U.S. military lexicon, dissuasion is said to work outside the potential threat of military action and has been called a kind of "pre-deterrence," or *deterrence by denial* using Glenn Snyder's terminology from 1960.[57] According to Snyder's definition, deterrence by denial is "the capability to deny the other party any gains from the move which is to deterred."[58] Drawing upon Snyder, Paul Davis defines the concept as *"deterring an action by having the adversary see a credible capability to prevent him from achieving potential gains adequate to motivate the action."*[59] A strategy incorporating dissuasion seeks to convey the futility of conducting a hostile act, thereby causing a potential adversary's leadership to not pursue a military confrontation in the first place.

A potential adversary may be dissuaded if it concludes that an attack in space will be ineffectual in achieving the desired effect. In the parlance of today's U.S. space professionals, this is the realm of space mission assurance. Space mission assurance efforts consist of: *defensive operations*, which includes off-board protection elements; *reconstitution*, which includes launching replacement satellites or activating new ground stations; and *resilience*, which includes on-board protection elements.[60] Of note, resilience includes disaggregation, distribution, and diversification. *Disaggregation* of capabilities is "the separation of dissimilar capabilities into separate platforms or payloads."[61] *Distribution* utilizes a number of nodes, working together, to perform the same mission or functions as a single node.[62] *Diversification* is contributing to the same mission in multiple ways, using different platforms, different orbits, or systems and capabilities of commercial, civil, or international partners.[63] In the end, space mission assurance may leverage cross-domain or alternative government, commercial, or international capabilities. Viable dissuasion measures include actions resulting in a potential adversary not seeking a military confrontation. Therefore, measures— including distribution, redundancy, maneuverability, and protection—are all appropriate for promoting dissuasion in space.

A key element of these examinations of dissuasion, or deterrence by denial, is the recognition that to dissuade aggression in space attacks, would-be aggressors must perceive that their attacks will be futile. This agrees with Everett Dolman's writings that significant defensive and offensive space capabilities may dissuade others from attempting to compete in space.[64] As with deterrence, any space mission assurance efforts, however, must be widely publicized to be effective in dissuading others. Mission assurance, inclusive of the idea of resilience, remains a primary means to affect a potential adversary's thinking when employing a deterrence by denial strategy.

The second aspect of dissuasion is having a reliable and responsive space forensics capacity to assist in the attribution process following a hostile act. As defined here, *space forensics* includes catalog information, along with data and signal analysis from satellites or ground systems, to help identify details of a hostile act. Space forensics, along with information from law enforcement and intelligence community sources, support the attribution process in assigning responsibility following a hostile act. A robust and effectively communicated capability to rapidly identify and attribute the source of attack in space or on critical ground segments may help dissuade potential adversaries who would only attack if their identity would remain unknown. Following a hostile act in space, a space forensics capability informing the attribution process may lead to prosecution through civilian courts, or for more significant acts of aggression, it may lead to targeting with kinetic or non-kinetic weapons against those attributed for the attack.

As with the requisites for preemption, a significant part of this forensics capability will be space situational awareness (SSA) capabilities, which are intended to provide knowledge of space objects and activities along with supporting timely attribution. SSA includes the requisite foundational, current, and predictive knowledge and characterization of space objects and the operational environment upon which space operations depend, as well as factors, activities and events of all entities conducting—or preparing to conduct—space operations.[65] These SSA capabilities may be either governmental or commercial in nature.

U.S. joint doctrine notes several of the previous points regarding attribution and SSA. In defining defensive space control (DSC), it is described that:

> DSC capabilities should be integrated with SSA elements that provide the ability to detect, characterize, and attribute an attack to an enemy. A robust DSC capability influences enemies' perceptions of US space capabilities and makes them less confident of success in interfering with those capabilities.[66]

Consequently, the ability to observe in real-time and understand what is happening in space leads to timely attribution and may affect the decisions of would-be adversaries.

The strategy mismatches

The strategist's job is to develop a practical strategy given the unique conditions in which it is to be implemented. Such a process is far from perfect. When formulating what would appear to be a logical and sound strategy, a time-tested adage must be remembered: "the enemy gets a vote." Framing the problem when considering deterrence, Steven Lambakis puts it poignantly, "Our values are not necessarily their values. Our ways may not be their ways. Just because we would not do it, does not mean they would not do it."[67] A strategy—including one in which deterrence is a central element—should only be judged as effective in relationship to how the strategy affects the mental calculus of another. Therefore, when considering deterrence by denial or punishment approaches, it is necessary to understand how a potential adversary's view may differ from one's own world view or implementation of a deterrence strategy.

Strategy mismatches—where there are different cultural and social understandings of deterrence and escalation control—are some of the most dangerous situations between states. This danger is because states, whose leaders may consider themselves to be rational and reasonable in not seeking direct military confrontation, may find themselves in such a war, despite their intent or desire. Because of the different understandings of deterrence in preventing war or deterrence's ability to control escalation during conflict, it is useful to contrast American and many Western countries' views against those considered "undeterrable," along with different deterrence definitions of Russia and China. The Russian military's strategy of "unacceptable losses" and the Chinese view of using "compellence" through military actions to avoid conflict are two different strategy approaches, which American policy-makers and strategists should understand well.

The undeterrable

When considering deterrence, it must be remembered that some people or foreign leaders will not be deterred. Putting the correct perspective on the efficacy of deterrence, Colin Gray observes, "Polities are not always deterrable; they may decline to be coerced, or even when heavily physically damaged, they may elect to soldier on and hope for a change in strategic fortune."[68] In reference to those who will not be swayed in their decision for violence regardless of the threat of a severe military response to a hostile attack, Gray calls such individuals "fools," because they are far more likely to commit errors of a kind that result in wars or at least a high measure of regional disorder.[69] He goes on to say, "Deterrence could be irrelevant in such a case, because the foolish foreign leader may not believe in the latent or explicit threats we issue, or, just possibly, may not care whether or not we execute them."[70] Karl Mueller has similarly noted, "if the enemy has nothing to lose, even a very risky action may be preferable" to maintaining the status quo.[71]

So, it does not matter whether one thinks a potential adversary should be deterred given an action or situation; it only matters how the adversary's leadership and decision-makers interpret any action within their world view and mental constructs. While such a situation may be disconcerting for those seeking to "guarantee deterrence," that is the reality of international affairs.

Russia

Over the last decade, Russia has been implementing its vision of strategic deterrence that is built on demonstrating a spectrum of capabilities and resolve to use military force. Russia's strategic deterrence is conceptually different from its Western namesake in that it is not limited to solely nuclear weapons.[72] In describing *strategic deterrence*, Russian military writings describe the term as an approach seeking to "induce fear" in opponents, whether in peace or war. Therefore, the concept includes elements of what others could call *deterrence*, *containment*, and *coercion*.[73] Russia's strategic deterrence approach is grounded in its understanding of internal and external threats, including a sense of military asymmetry compared to the West.[74] Russian military doctrine describes perceived dangers from the United States and NATO readiness to use military force, instability and terrorism that could challenge Russia's sovereignty, and a local conflict on its vast borders that could escalate into hostilities, which could include the use of nuclear weapons.[75]

In the Russian perspective, strategic deterrence is not entirely defensive. Within U.S. security circles, some may consider Russia's view of strategic deterrence as an "escalate to deescalate" strategy—even though that term is not used within Russian military doctrine or strategies—because the idea includes using military force and actions to potentially deescalate hostilities or tensions.[76] The Russian concept transcends a traditional perception of deterrence having failed if conflict erupts. Therefore, deterrence can continue to work "in times of war to prevent escalation, to ensure de-escalation, or for the swift termination of conflict on terms acceptable to Russia."[77] Strategic deterrence seeks to influence wartime calculations through demonstrating Russian willingness to use coercive measures. Whereas the sheer destructiveness of nuclear weapons means their mere existence should be enough to deter, it is thought that non-nuclear and non-military measures, in particular, must be demonstrated or used coercively to deter a potential adversary.[78] The Russian term *strategic deterrence* is thus an inclusive concept describing: activities aimed at preventing any threat from materializing against Russia; activities aimed at deterring any direct aggression against Russia; and, last, activities focused on coercing an adversary to cede in a confrontation to terms dictated by Russia.[79]

Besides the large scale use of strategic nuclear weapons—which is considered to inflict "deterrent damage"—the threat of limited or non-strategic nuclear weapon use is also thought to have a deterrent effect.[80] Limited use of nuclear weapons could de-escalate and terminate combat actions on terms acceptable to Russia through the threat of inflicting "unacceptable damage" upon the enemy.[81] Consequently, limited use of nuclear weapons is thought to deter both nuclear and conventional aggression. Although many Western analysts may assume that non-strategic nuclear weapons are the most likely option for such limited use, most Russian analysts make no distinction between strategic or sub-strategic nuclear weapons in this respect.[82]

Additionally, Russian doctrine describes the threat of the massive use of non-strategic nuclear forces and strategic non-nuclear forces, under the idea of *regional deterrence*, the result of which might include the destruction of the opposing military forces and irreparable damage to the economy of the aggressor. Emphasis on the interchangeability of conventional precision weapons and limited or non-strategic nuclear weapons is habitual within Russian doctrine. Current Russian thinking is that conventional weapons could carry out missions like those of nuclear weapons, such as demonstration strikes and limited strikes aimed at de-escalation, while also destroying targets critical to the enemy.[83]

Russia's strategic deterrence concept highlights that a misunderstanding regarding intent could well fuel escalation dynamics, especially with those holding to a Western view of deterrence.[84] In a nascent crisis, it is thought that Russia is likely to engage in deterrence signaling and increase the readiness of selected conventional and perhaps nuclear capabilities. Most notably, Russia's plans to control escalation by using conventional precision-strike missile systems on an opponent's military and economic targets increases the likelihood of unintended escalation, especially when employed alongside cyber and electronic warfare attacks.[85] Communicating what actions may result in retaliation constitutes a key element of deterrence strategy, but Russia's expanded deterrence concept is noted to be deficient in this regard. While Russia's strategic deterrence seeks to exploit the attention and fear generated by indirect uses of military force, Russian analysts have also argued that Moscow must seriously engage Western proposals on transparency of conventional forces.[86]

China

As with Russia, the Chinese concept of deterrence is fundamentally different than American and Western thinking. In their analysis, Alison Kaufman and Daniel Hartnett note the Chinese concept of deterrence (*weishe*) includes a significant element of compellence and coercion; therefore, Chinese deterrence goals may include actions seeking to intimidate the opponent through economic, diplomatic, or military coercion in a way that "directly affect[s] an opponent's interests in order to compel him to submit to Beijing's will."[87] In the 2001 edition of the Science of Military Strategy, the dual nature of this idea was highlighted in defining *strategic deterrence* as "a military strategy [in which one] displays or threatens to use force in order to compel (*poshi*) the adversary to yield."[88] Analysts of Chinese strategy urge readers to also keep in mind the nuances in the terms used, especially those with more coercive connotations.[89] Dean Cheng has similarly noted when describing the difference between Chinese and American views of deterrence, "The Chinese focus is on compellence, including coercion, rather than solely, or even primarily, on dissuasion. Thus, the idea of 'deterrence' is seen in both coercive and dissuasive terms."[90] As a result, the Chinese see deterrence as a means to achieving political ends.[91]

More importantly, the phases of crisis and conflict differ between the United States and China. According to analysts, Chinese writings consistently identify a continuum of conflict by describing a series of stages in the progression from least to greatest crisis and conflict. These stages across the continuum are: "crisis; military crisis; armed conflict; local war; total war."[92] The most potentially dangerous state on the continuum of conflict is thought to be the middle part of the continuum, in which military activities are taking place and the objectives are less clear. This middle of the continuum includes *military crisis* and/or *armed conflict*, in which militaries are involved but war has not yet broken out.[93] Military operations in the state of "quasi-war" appear to have dual objectives. The first is to resolve the crisis and prevent the onset of war, and the second is to prepare to win a war should one break out.[94] Several People's Liberation Army (PLA) texts argue that during a state of pre-war "armed conflict," countries may take limited military action to "clarify the situation" or persuade the other side to de-escalate.[95] According to PLA writings, military activities in this stage may resemble combat operations, even if the countries involved do not consider themselves to be at war. Of concern is that PLA writings do not provide any clear indications of how an outside observer would discern the differing intentions of these military operations.

Another difference in thinking relates to deterrence during war versus deterrence in each domain of warfare. China does not appear interested in "deterrence in space," or deterring an adversary from acting in the space domain or acting against space assets. Deterrence is thought of holistically and not isolated to each domain of potential conflict. Instead, China's strategists are focused on "deterrence through space," thereby integrating space activities with conventional, cyber, and even nuclear to influence an adversary.[96] Additionally, Dean Cheng has observed, regarding the Chinese view of space deterrence's broad impact, "This reinforces the point that, from the Chinese perspective, 'space deterrence' is not about deterring adversaries from acting in space, but exploiting space-related systems to achieve certain political and military aims (largely on Earth)."[97]

Alison Kaufman and Daniel Hartnett are concerned, because it is unclear whether U.S. leadership, policy-makers, and strategists grasp the important distinctions between Chinese and American views.[98] PLA writings promote several crisis and conflict control actions that could appear escalatory. In combination, the PLA notion that there can be a stage of armed conflict short of war—together with a doctrine that advocates going on the offensive early in a war—has serious escalatory implications.[99] The 2013 Science of Military Strategy says that it is important to "not be afraid to (*ganyu*) use military deterrence methods, particularly in space, network and other new domains of struggle, to smash the enemy's warfighting command systems."[100] Any of these could be perceived by an opponent as escalatory if initiated during a crisis—even if the PLA does not intend them to be perceived as such.[101] As a result, Kaufman and Hartnett warn there is a high likelihood of misperception and misunderstanding between China and the United States in the state of "quasi-war."[102] Because of the PLA's well-known emphasis on seizing the initiative in war, one can envision a situation where the PLA takes what it intends to be a limited military action in a state of pre-war but an adversary assumes that it is the beginning of a large-scale attack.[103]

Conclusion

The strategy of space warfare is a subset of general warfare strategy. Consequently, the ideas of the inherent right of self-defense, deterrence, preemption, and dissuasion have applicability in space strategy. Even though deterrence has a legitimate role in future space strategy, it is not the panacea for preventing conflict. History teaches that deterrence will at times fail due to miscalculation, uncertainty, or chance—ideas incorporating the concept of Clausewitzian friction. This may also be the case for deterring acts of aggression in space, especially considering countries like the United States, Russia, and China have different perspectives on deterrence, compellence, and escalation control.

Anyone believing they have the most to lose in a conflict extending into space—to include some security experts in the United States—should consider incorporating anticipatory self-defense into their contingency planning. This is especially the case given the likelihood of potential strategy mismatches between competing states. Albeit the topic of preemption is a contentious issue with many security professionals, customary international law has long supported the view that a state can attack first in a proportional manner, when under imminent threat of armed attack or hostile act.

Practical strategies incorporating deterrence, preemption, and dissuasion require exceptional SSA, space forensics, and resilience capabilities, along with a timely space attribution process. Additionally, these concepts require more progress by the legal community in advancing what constitutes *hostile intent, hostile act,* and *armed attack* in the space domain, as well as incorporating such definitions into Rules of Engagement. These efforts also necessitate communicating with the international community regarding these definitions and under what conditions self-defense and anticipatory self-defense are warranted.

Finally, more dialogue and debate regarding deterrence, dissuasion, and anticipatory self-defense should be welcomed among the security and policy communities, even if it results in the view that a country presently lacks the requisite capabilities and processes. Space powers need to be able to respond at any point within a space conflict timeline and at any location, whether preemptively or after being attacked. Such dialogue—including any resulting improvement to space strategies and capabilities—helps promote international peace and stability, while also helping ensure national interests in space are better protected.

Notes

1 Thomas C. Schelling, *Arms and Influence* (New Haven, CT: Yale University Press, 1966), 2, 31–34.
2 Ibid., x.
3 Ibid., 2.
4 Ibid., 72.
5 Ibid., 69–72.
6 "Effective deterrence also requires a consistency and clarity in signals and messaging...." Joan Johnson-Freese, *Space Warfare in the 21st Century: Arming the Heavens* (Abingdon: Routledge, 2017), xii.
7 James P. Finch and Shawn Steene, "Finding Space in Deterrence: Toward a General Framework for Space Deterrence," *Strategic Studies Quarterly* vol. 5 no. 4 (Winter 2011), 13, www.dtic.mil/dtic/tr/fulltext/u2/a569581.pdf
8 Ibid.
9 James A. Vedda and Peter L. Hays, "Major Policy Issues in Evolving Global Space Operations," (The Mitchell Institute of Aerospace Studies, February 2018), 3, www.aerospace.org/publications/policy-papers/major-policy-issues-in-evolving-global-space-operations/
10 Bernard Brodie, "The Development of Nuclear Strategy," *International Security* vol. 2 no. 4 (Spring, 1978), 65–83.
11 Finch and Steene, "Finding Space in Deterrence," 10–17.
12 Karl Mueller, "The Absolute Weapon and the Ultimate High Ground: Why Nuclear Deterrence and Space Deterrence Are Strikingly Similar – Yet Profoundly Different," in *Anti-satellite Weapons, Deterrence and Sino-American Space Relations,* eds. Michael Krepon and Julia Thompson (The Stimson Center, September 2013), 41–60, www.stimson.org/content/anti-satellite-weapons-deterrence-and-sino-American-space-relations

13 James A. Lewis, "Reconsidering Deterrence for Space and Cyberspace" in *Anti-satellite Weapons, Deterrence and Sino-American Space Relations*, eds. Michael Krepon and Julia Thompson (The Stimson Center, September 2013), 61–80.

14 Michael Krepon, "Space and Nuclear Deterrence," in *Anti-satellite Weapons, Deterrence and Sino-American Space Relations*, eds. Michael Krepon and Julia Thompson (The Stimson Center, September 2013), 15.

15 Ibid.

16 Ibid.

17 Ibid.

18 Ibid., 38.

19 Department of Defense, *U.S. Quadrennial Defense Review* (September 30, 2001), 14, http://archive.defense.gov/pubs/qdr2001.pdf

20 David J. Trachtenberg, "US Extended Deterrence How Much Strategic Force Is Too Little?" *Strategic Studies Quarterly* vol. 6 no. 2 (Summer 2012); www.airuniversity. af.mil/Portals/10/SSQ/documents/Volume-06_Issue-2/05-Trachtenberg.pdf

21 Colin S. Gray, *Weapons Don't Make War: Policy, Strategy, and Military Technology* (Lawrence, KS: University Press of Kansas, 1993), 25.

22 Dean Cheng, "Prospects for Extended Deterrence in Space and Cyber: The Case of the PRC," Lecture No. 1270 (The Heritage Foundation, January 21, 2016), www. heritage.org/research/reports/2016/01/prospects-for-extended-deterrence-in-space-and-cyber-the-case-of-the-prc

23 Wing Commander Steve Henry, Australian Air Force, "US Allies in Space Operations" (presentation, The Mitchell Institute for Aerospace Studies, Washington, DC, October 27, 2017).

24 Ibid.

25 Ibid.

26 Gregory Schulte, "Protecting NATO's Advantage in Space," *Transatlantic Current* No. 5 (National Defense University, May 2012). www.dtic.mil/dtic/tr/fulltext/u2/ a577645.pdf

27 Ibid.

28 Article 51 states,

> Nothing in this present Charter shall impair the inherent right of individual or collective self-defense if an armed attack occurs against a Member of the United Nations, until the Security Council has taken measures necessary to maintain international peace and security.
>
> United Nations, *Charter of the United Nations and Statue of the International Court of Justice* (San Francisco, June 26, 1945), Chapter 7, Article 51

29 Schelling, *Arms and Influence*, 35.

30 Joint Chiefs of Staff, *Department of Defense Dictionary of Military and Associated Terms*, Joint Publication 1–02 (March 23, 1994), 215.

31 Ibid.

32 Note that *armed attack* is the phase used in the United Nations Charter, Article 51. Ibid.

33 U.S. Department of the Navy, *The Commander's Handbook on the Law of Naval Operations*, NWP 1–14M (July 9, 1995), 6–5.

34 Ibid., 8–1.

35 Ibid., 6–5.

36 Joint Chiefs of Staff, *Standing Rules of Engagement for US Forces*, CJCSI 3121.01A (January 15, 2000), 1.

37 Ibid., (Enclosure A), p. A-4.

38 General John Hyten, "Space, Nuclear, and Missile Defense Modernization" (presentation, The Mitchell Institute for Aerospace Studies, Washington, DC, June 20,

2017), www.stratcom.mil/Media/Speeches/Article/1226883/mitchell-institute-breakfast-series/

39 United Nations, *Charter of the United Nations*, Chapter 1, Article 2(4).

40 Chairman of the Joint Chiefs of Staff Instruction, "Standing Rules of Engagement," in *Operational Law Handbook*, 16th edition, eds. Rachel Mangas and Matthew Festa (Charlottesville, VA: The Judge Advocate General's Legal Center and School, U.S. Army, 2016), 95–110, www.loc.gov/rr/frd/Military_Law/pdf/OLH_2015_Ch5.pdf

41 General John Hyten in *60 Minutes*, "The Battle Above," David Martin, original air date, August 2, 2015, www.cbsnews.com/news/rare-look-at-space-command-satellite-defense-60-minutes-2/

42 Colin S. Gray, "The Implications of Preemption and Preventive War Doctrines: A Reconsideration" (Strategic Studies Institute, July 2007), v, http://ssi.armywar college.edu/pdffiles/pub789.pdf

43 United Nations, *Charter of the United Nations*, Chapter 7, Article 51.

44 Anthony Clark Aren, "International Law and the Preemptive Use of Military Force," *The Washington Quarterly*, vol. 26 no. 2 (Spring 2003), 89–103, www.cfr.org/content/publications/attachments/highlight/03spring_arend.pdf

45 The White House, *The National Security Strategy of the United States of America* (September 2002), 15, www.state.gov/documents/organization/63562.pdf

46 Edward G. Ferguson and John J. Klein, "It's Time for the U.S. Air Force to Prepare for Preemption in Space," *War is Boring*, April 22, 2017, https://warisboring.com/its-time-for-the-u-s-air-force-to-prepare-for-preemption-in-space/

47 James Fallows, "John Boyd in the News: All You Need to Know About the OODA Loop," *The Atlantic*, August 29, 2015, www.theatlantic.com/notes/2015/08/john-boyd-in-the-news-all-you-need-to-know-about-ooda-loop/402847/

48 Jay Bennett, "Space War: How the Air Force Plans to Defend the Final Frontier," *Popular Mechanics*, November 6, 2017, www.popularmechanics.com/military/news/a28851/us-air-force-gears-up-for-expanded-role-in-space/

49 Ibid.

50 Eldbridge Colby, "From Sanctuary to Battlefield: A Framework for a U.S. Defense and Deterrence Strategy for Space" (Center for a New American Security, January 2016), 11, www.cnas.org/publications/reports/from-sanctuary-to-battlefield-a-framework-for-a-us-defense-and-deterrence-strategy-for-space

51 Brian G. Chow, "Stalkers in Space: Defeating the Threat," *Strategic Studies Quarterly* vol. 11 no. 2 (Summer 2017), 100.

52 Ibid.

53 Gray, "The Implications of Preemption and Preventive War Doctrines," v.

54 Ferguson and Klein, "It's Time for the U.S. Air Force to Prepare for Preemption in Space."

55 Department of Defense, *U.S. Quadrennial Defense Review* (September 30, 2001).

56 Glen M. Segall, "Thoughts on Dissuasion," *Journal of Military and Strategic Studies* vol. 10 no. 4 (Summer 2008), 1, https://jmss.org/article/view/57658/43328

57 Andrew F. Krepinevich and Robert C. Martinage, "Dissuasion Strategy" (Center for Strategic and Budgetary Assessments, 2008), www.files.ethz.ch/isn/162559/2008.05.06-Dissuasion-Strategy.pdf; Glenn Snyder, "Deterrence and Power," *Journal of Conflict Resolution* vol. 4 no. 2 (June 1960), 163–178.

58 Snyder, "Deterrence and Power," 163.

59 Paul K. Davis, "Toward Theory for Dissuasion (or Deterrence) by Denial: Using Simple Cognitive Models of the Adversary to Inform Strategy," RAND NSRD WR-1027 (RAND Corporation, January 2014) 2, Emphasis original. www.rand.org/content/dam/rand/pubs/working_papers/WR1000/WR1027/RAND_WR1027.pdf

60 Office of the Assistant Secretary of Defense for Homeland Defense and Global Security, *Space Domain Mission Assurance: A Resilience Taxonomy* (September 2015), 3, http://policy.defense.gov/Portals/11/Space%20Policy/ResilienceTaxonomy WhitePaperFinal.pdf?ver=2016-12-27-131828-623

61 Office of the Assistant Secretary of Defense for Homeland Defense and Global Security, *Space Domain Mission Assurance*, 6.

62 Ibid.

63 Ibid., 7.

64 Everett C. Dolman, *Astropolitik: Classical Geopolitics in the Space Age* (London: Frank Cass, 2002), 156–158.

65 Joint Chiefs of Staff, *Space Operations*, Joint Publication 3–14 (10 April, 2018), II–2. www.jcs.mil/Portals/36/Documents/Doctrine/pubs/jp3_14.pdf

66 Joint Chiefs of Staff, *Space Operations*, II–2.

67 Steven Lambakis, *On the Edge of Earth: The Future of American Space Power* (Lexington, KY: The University Press of Kentucky, 2001), 183.

68 Colin S. Gray, *Airpower for Strategic Effect* (Maxwell Air Force Base, AL: Air University Press, 2012), 296.

69 Colin S. Gray, *Fighting Talk: Forty Maxims on War, Peace, and Strategy* (Westport, CT: Greenwood Publishing, 2007), 125.

70 Ibid.

71 Mueller, "The Absolute Weapon and the Ultimate High Ground," 43.

72 Anya Loukianova Fink, "The Evolving Russian Concept of Strategic Deterrence: Risks and Responses," Arms Control Today (Arms Control Association, July/August 2017), www.armscontrol.org/act/2017–07/features/evolving-russian-concept-strategic-deterrence-risks-responses

73 Yu A. Pechatnov, "Deterrence Theory: Beginnings," *Vooruzheniye I Ekonomika* (February 2016). As referenced in Fink, "The Evolving Russian Concept of Strategic Deterrence."

74 Ibid.

75 Embassy of the Russian Federation to the United Kingdom of Great Britain and Northern Ireland, "The Military Doctrine of the Russian Federation" (June 29, 2015), http://rusemb.org.uk/press/2029

76 Mark B. Schneider, "Escalate to De-escalate," U.S. Naval Institute *Proceedings* vol. 143/2/1,368 (February 2017), www.usni.org/magazines/proceedings/2017–02/escalate-de-escalate

77 Russian Federation Ministry of Defense, "Military-Encyclopedic Dictionary of the Russian Ministry of Defense," accessed August 11, 2018, http://encyclopedia.mil.ru/encyclopedia/dictionary/details.htm?id=14206@morfDictionary. As referenced in Kristin Ven Bruusgaard, "Russian Strategic Deterrence," *Survival: Global Politics and Strategy* vol. 58 no. 4 (August–September 2016), 7–26, www.iiss.org/en/publications/survival/sections/2016–5e13/survival–global-politics-and-strategy-august-september-2016–2d3c/58–4-02-ven-bruusgaard-45ec

78 Dmitry (Dima) Adamsky, "Cross-Domain Coercion: The Current Russian Art of Strategy," *Proliferation Papers* vol. 54 (2015). As referenced in Ven Bruusgaard, "Russian Strategic Deterrence."

79 Ibid.

80 D.A. Kalinkin, A.L. Khryapin, and V.V. Matvichuk, "Strategic Deterrence in the Context of the US Global Ballistic-Missile Defense System and Means for Global Strike," *Voyennaya Mysl'* no. 1 (January 2015), 18–22. As referenced in Ven Bruusgaard, "Russian Strategic Deterrence."

81 See S.A. Bogdanov and S.G. Chekinov, "Strategic Deterrence and Russian National Security in the Contemporary Era," *Voyennaya Mysl'* no. 3 (March 2012), 11–20. As referenced in Ven Bruusgaard, "Russian Strategic Deterrence."

82 Ven Bruusgaard, "Russian Strategic Deterrence," 7–26.

83 V.A. Sobolevskii', A.A. Protasov and V.V. Sukhorutchenko, "Planning for the Use of Strategic Weapons," *Voyennaya Mysl'* no. 7 (July 2014), 9–27. As referenced in Ven Bruusgaard, "Russian Strategic Deterrence," 7–26.

84 Fink, "The Evolving Russian Concept of Strategic Deterrence."

85 Ibid.
86 Sergei Oznobishchev, "Russia and NATO: From the Ukrainian Crisis to the Renewed Interaction," in "Russia: Arms Control, Disarmament, and International Security," eds. Alexei Arbatov and Sergei Oznobishchev (Moscow: IMEMO, 2016), www.sipri.org/sites/default/files/SIPRI-Yearbook-Supplement-2015.pdf
87 Mark A. Stokes, "The Chinese Joint Aerospace Campaign: Strategy, Doctrine, and Force Modernization," in "China's Revolution in Doctrinal Affairs: Emerging Trends in the Operational Art of the Chinese People's Liberation Army," eds. James Mulvenon and David M. Finkelstein (CNA, 2005), 226–227. Referenced in Alison A. Kaufman and Daniel M. Hartnett, "Managing Conflict: Examining Recent PLA Writings on Escalation Control" (CNA, February 2016), 54.
88 Peng Guangqian and Yao Youzhi, eds., *The Science of Military Strategy* (Beijing: Military Science Publishing House, 2001), 230. Referenced in Kaufman and Hartnett, "Managing Conflict: Examining Recent PLA Writings on Escalation Control," 54.
89 Kaufman and Hartnett, "Managing Conflict: Examining Recent PLA Writings on Escalation Control," 54.
90 Dean Cheng, "Evolving Chinese Thinking about Deterrence: What the United States Must Understand About China and Space," Backgrounder No. 3298 (The Heritage Foundation, March 29, 2018), 2, http://report.heritage.org/bg3298
91 Ibid.
92 Kaufman and Hartnett, "Managing Conflict: Examining Recent PLA Writings on Escalation Control," 20.
93 Ibid.
94 Ibid.
95 Ibid.
96 Ibid.
97 Cheng, "Evolving Chinese Thinking About Deterrence: What the United States Must Understand About China and Space," 2.
98 Dean Cheng, "Are We Ready to Meet the Chinese Space Challenge?" *Spacenews*, July 10, 2017, http://spacenews.com/op-ed-are-we-ready-to-meet-the-chinese-space-challenge/
99 Kaufman and Hartnett, "Managing Conflict: Examining Recent PLA Writings on Escalation Control," v.
100 Shou Xiaosong, ed., *The Science of Military Strategy* (Beijing: Military Science Press, 2013), 129. Referenced in Kaufman and Hartnett, "Managing Conflict: Examining Recent PLA Writings on Escalation Control," 56.
101 Ibid.
102 Ibid.
103 Ibid.

Bibliography

60 Minutes. "The Battle Above." David Martin. Original air date, August 2, 2015. www.cbsnews.com/news/rare-look-at-space-command-satellite-defense-60-minutes-2/

Adamsky, Dmitry (Dima). "Cross-Domain Coercion: The Current Russian Art of Strategy." *Proliferation Papers* vol. 54 (2015).

Aren, Anthony Clark. "International Law and the Preemptive Use of Military Force." *The Washington Quarterly* vol. 26 no. 2 (Spring 2003): 89–103. www.cfr.org/content/publications/attachments/highlight/03spring_arend.pdf

Bennett, Jay. "Space War: How the Air Force Plans to Defend the Final Frontier." *Popular Mechanics*. November 6, 2017. www.popularmechanics.com/military/news/a28851/us-air-force-gears-up-for-expanded-role-in-space/

Bogdanov S.A. and S.G. Chekinov. "Strategic Deterrence and Russian National Security in the Contemporary Era." *Voyennaya Mysl'* no. 3 (March 2012): 11–20.

Brodie, Bernard. "The Development of Nuclear Strategy." *International Security* vol. 2 no. 4 (Spring, 1978): 65–83.

Chairman of the Joint Chiefs of Staff Instruction. "Standing Rules of Engagement." In *Operational Law Handbook.* 16th edition. Edited by Rachel Mangas and Matthew Festa, 95–110. Charlottesville, VA: The Judge Advocate General's Legal Center and School, U.S. Army, 2016. www.loc.gov/rr/frd/Military_Law/pdf/OLH_2015_Ch5.pdf

Cheng, Dean. "Are We Ready to Meet the Chinese Space Challenge?" *Spacenews.* July 10, 2017. http://spacenews.com/op-ed-are-we-ready-to-meet-the-chinese-space-challenge

Cheng, Dean. "Evolving Chinese Thinking about Deterrence: What the United States Must Understand about China and Space." Backgrounder No. 3298. The Heritage Foundation, March 29, 2018. http://report.heritage.org/bg3298

Cheng, Dean. "Prospects for Extended Deterrence in Space and Cyber: The Case of the PRC." Lecture No. 1270. The Heritage Foundation, January 21, 2016. www.heritage.org/research/reports/2016/01/prospects-for-extended-deterrence-in-space-and-cyber-the-case-of-the-prc

Chow, Brian G. "Stalkers in Space: Defeating the Threat." *Strategic Studies Quarterly* vol. 11 no. 2 (Summer 2017): 82–116.

Colby, Eldbridge. "From Sanctuary to Battlefield: A Framework for a U.S. Defense and Deterrence Strategy for Space." Center for a New American Security, January 2016. www.cnas.org/publications/reports/from-sanctuary-to-battlefield-a-framework-for-a-us-defense-and-deterrence-strategy-for-space

Davis, Paul K. "Toward Theory for Dissuasion (or Deterrence) by Denial: Using Simple Cognitive Models of the Adversary to Inform Strategy." RAND NSRD WR-1027. RAND Corporation, January 2014. www.rand.org/content/dam/rand/pubs/working_papers/WR1000/WR1027/RAND_WR1027.pdf

Department of Defense. *U.S. Quadrennial Defense Review.* September 30, 2001. http://archive.defense.gov/pubs/qdr2001.pdf

Dolman, Everett C. *Astropolitik: Classical Geopolitics in the Space Age.* London: Frank Cass, 2002.

Embassy of the Russian Federation to the United Kingdom of Great Britain and Northern Ireland. "The Military Doctrine of the Russian Federation." June 29, 2015. http://rusemb.org.uk/press/2029

Fallows, James. "John Boyd in the News: All You Need to Know About the OODA Loop." *The Atlantic.* August 29, 2015. www.theatlantic.com/notes/2015/08/john-boyd-in-the-news-all-you-need-to-know-about-ooda-loop/402847/

Ferguson, Edward G. and John J. Klein. "It's Time for the U.S. Air Force to Prepare for Preemption in Space." *War is Boring.* April 22, 2017, https://warisboring.com/its-time-for-the-u-s-air-force-to-prepare-for-preemption-in-space/

Finch, James P. and Shawn Steene. "Finding Space in Deterrence: Toward a General Framework for Space Deterrence." *Strategic Studies Quarterly* vol. 5 no. 4 (Winter 2011): 10–17. www.dtic.mil/dtic/tr/fulltext/u2/a569581.pdf

Fink, Anya Loukianova. "The Evolving Russian Concept of Strategic Deterrence: Risks and Responses." Arms Control Today. Arms Control Association, July/August 2017. www.armscontrol.org/act/2017-07/features/evolving-russian-concept-strategic-deterrence-risks-responses

Gray, Colin S. "The Implications of Preemption and Preventive War Doctrines: A Reconsideration." Strategic Studies Institute, July 2007. http://ssi.armywarcollege.edu/pdffiles/pub789.pdf

Gray, Colin S. *Airpower for Strategic Effect*. Maxwell Air Force Base, AL: Air University Press, 2012.

Gray, Colin S. *Fighting Talk: Forty Maxims on War, Peace, and Strategy*. Westport, CT: Greenwood Publishing, 2007.

Gray, Colin S. *Weapons Don't Make War: Policy, Strategy, and Military Technology*. Lawrence, KS: University Press of Kansas, 1993.

Henry, Steve. "US Allies in Space Operations," Presentation. The Mitchell Institute for Aerospace Studies, Washington, DC, October 27, 2017.

Hyten, John. "Space, Nuclear, and Missile Defense Modernization." Presentation. The Mitchell Institute for Aerospace Studies, Washington, DC, June 20, 2017. www.stratcom.mil/Media/Speeches/Article/1226883/mitchell-institute-breakfast-series/

Johnson-Freese, Joan. *Space Warfare in the 21st Century: Arming the Heavens*. Abingdon: Routledge, 2017.

Joint Chiefs of Staff. *Department of Defense Dictionary of Military and Associated Terms*. Joint Publication 1–02. March 23, 1994.

Joint Chiefs of Staff. *Space Operations*. Joint Publication 3–14. April 10, 2018. www.jcs.mil/Portals/36/Documents/Doctrine/pubs/jp3_14.pdf.

Joint Chiefs of Staff. *Standing Rules of Engagement for US Forces*. CJCSI 3121.01A. January 15, 2000.

Kalinkin, D.A., A.L. Khryapin, and V.V. Matvichuk, "Strategic Deterrence in the Context of the US Global Ballistic-Missile Defense System and Means for Global Strike," *Voyennaya Mysl'* no. 1 (January 2015): 18–22.

Kaufman, Alison A. and Daniel M. Hartnett. "Managing Conflict: Examining Recent PLA Writings on Escalation Control." CNA, February 2016.

Krepinevich, Andrew F. and Robert C. Martinage. "Dissuasion Strategy." Center for Strategic and Budgetary Assessments, 2008. www.files.ethz.ch/isn/162559/2008.05.06-Dissuasion-Strategy.pdf

Krepon, Michael. "Space and Nuclear Deterrence." In *Anti-satellite Weapons, Deterrence and Sino-American Space Relations*, edited by Michael Krepon and Julia Thompson, 15–40. The Stimson Center, September 2013. www.stimson.org/content/anti-satellite-weapons-deterrence-and-sino-American-space-relations

Lambakis, Steven. *On the Edge of Earth: The Future of American Space Power*. Lexington, KY: The University Press of Kentucky, 2001.

Lewis, James A. "Reconsidering Deterrence for Space and Cyberspace." In *Anti-satellite Weapons, Deterrence and Sino-American Space Relations*, edited by Michael Krepon and Julia Thompson, 61–80. The Stimson Center, September 2013. www.stimson.org/content/anti-satellite-weapons-deterrence-and-sino-American-space-relations

Mueller, Karl. "The Absolute Weapon and the Ultimate High Ground: Why Nuclear Deterrence and Space Deterrence Are Strikingly Similar – Yet Profoundly Different." In *Anti-satellite Weapons, Deterrence and Sino-American Space Relations*, edited by Michael Krepon and Julia Thompson, 41–60. The Stimson Center, September 2013. www.stimson.org/content/anti-satellite-weapons-deterrence-and-sino-American-space-relations

Office of the Assistant Secretary of Defense for Homeland Defense and Global Security. *Space Domain Mission Assurance: A Resilience Taxonomy*. September 2015. http://policy.defense.gov/Portals/11/Space%20Policy/ResilienceTaxonomyWhitePaperFinal.pdf?ver=2016-12-27-131828-623

Oznobishchev, Sergei. "Russia and NATO: From the Ukrainian Crisis to the Renewed Interaction." In *Russia: Arms Control, Disarmament, and International Security*, edited by Alexei Arbatov and Sergei Oznobishchev, 57–71. Moscow: IMEMO, 2016. www. sipri.org/sites/default/files/SIPRI-Yearbook-Supplement-2015.pdf

Pechatnov, Yu. A. "Deterrence Theory: Beginnings." *Vooruzheniye I Ekonomika* (February 2016).

Peng Guangqian and Yao Youzhi, eds. *The Science of Military Strategy.* Beijing: Military Science Publishing House, 2001.

Russian Federation Ministry of Defense. "Military-Encyclopedic Dictionary of the Russian Ministry of Defense." Accessed August 11, 2018. http://encyclopedia.mil.ru/ encyclopedia/dictionary/details.htm?id=14206@morfDictionary

Schelling, Thomas C. *Arms and Influence.* New Haven, CT: Yale University Press, 1966.

Schneider, Mark B. "Escalate to De-escalate." U.S. Naval Institute *Proceedings* vol. 143/2/1,368 (February 2017). www.usni.org/magazines/proceedings/2017-02/escalate- de-escalate

Schulte, Gregory. "Protecting NATO's Advantage in Space." *Transatlantic Current* No. 5. National Defense University, May 2012. www.dtic.mil/dtic/tr/fulltext/u2/a577645.pdf

Segall, Glen M. "Thoughts on Dissuasion." *Journal of Military and Strategic Studies* vol. 10 no. 4 (Summer 2008). https://jmss.org/article/view/57658/43328

Shou Xiaosong, ed. *The Science of Military Strategy.* Beijing: Military Science Press, 2013.

Snyder, Glenn. "Deterrence and Power." *Journal of Conflict Resolution* vol. 4 no. 2 (June 1960): 163–178.

Sobolevskii', V.A., A.A. Protasov and V.V. Sukhorutchenko. "Planning for the Use of Strategic Weapons." *Voyennaya Mysl'* no. 7 (July 2014): 9–27.

Stokes, Mark A. "The Chinese Joint Aerospace Campaign: Strategy, Doctrine, and Force Modernization." In "China's Revolution in Doctrinal Affairs: Emerging Trends in the Operational Art of the Chinese People's Liberation Army," edited by James Mulvenon and David M. Finkelstein, 221–305. CNA, 2005.

The White House. *The National Security Strategy of the United States of America.* September 2002. www.state.gov/documents/organization/63562.pdf

Trachtenberg, David J. "US Extended Deterrence How Much Strategic Force Is Too Little?" *Strategic Studies Quarterly* vol. 6 no. 2 (Summer 2012): 62–92. www.air university.af.mil/Portals/10/SSQ/documents/Volume-06_Issue-2/05-Trachtenberg.pdf

U.S. Department of the Navy. *The Commander's Handbook on the Law of Naval Operations.* NWP 1–14M. July 9, 1995.

United Nations. *Charter of the United Nations and Statue of the International Court of Justice.* San Francisco, June 26, 1945.

Vedda, James A. and Peter L. Hays. "Major Policy Issues in Evolving Global Space Operations." The Mitchell Institute of Aerospace Studies, February 2018. www. aerospace.org/publications/policy-papers/major-policy-issues-in-evolving-global- space-operations/

Ven Bruusgaard, Kristin. "Russian Strategic Deterrence." *Survival: Global Politics and Strategy* vol. 58 no. 4 (August–September 2016): 7–26, www.iiss.org/en/publications/ survival/sections/2016-5e13/survival-global-politics-and-strategy-august-september- 2016-2d3c/58-4-02-ven-bruusgaard-45ec

5 Space strategy for great powers

Historical experience illustrates that states will compete in space, and this competition includes an assessment of fear, honor, and interest. When considering the space strategies for states, the categories of *great*, *medium*, and *emerging powers* are useful for discerning relevant considerations for strategy's formulation. Even though the concepts and principles described in Chapter 2 stand on their own, the strategies for states and actors with varying degrees of power and capabilities will likely be different by necessity. Strategy involves balancing desired ends with available means, and the available means will be predicated, in part, on space-related capabilities commensurate with each category of space power. Because of the expected strategy preferences among the different types of space powers, the next three chapters consider those areas likely to be most germane for great, medium, and emerging space powers. That said, just because a concept is described in one category of power does not mean that it cannot be thoughtfully considered and implemented in another. Every competition or conflict between states is different, and therefore, the concepts discussed here are not meant to be prescriptive but only illustrative of those areas where thoughtful deliberations should occur.

Deganit Paikowsky has provided an insightful arrangement for those considered to be in the "space club," which can be used to differentiate between the levels of emerging, medium, and great space powers.[1] Using Paikowsky's framework, emerging space powers include the numerous states that indigenously can develop, maintain and control satellites but that are unable to launch them through indigenous means. There are many countries within the group, but examples include Canada and Saudi Arabia. Medium space powers include those states with the indigenous capability to launch, develop, and control satellites.[2] The contenders presently for the label of medium space power include the European Space Agency (ESA), the European member states of ESA that support ESA's space launch capabilities, Japan, India, Israel, Ukraine (which inherited its launch capability through the former Soviet Union), and Iran. In contrast, great space powers are defined as those having the aforementioned medium space power capabilities but also having the indigenous capability of human spaceflight, which includes China, Russia, and the United States. Even though the United States has not had indigenous human spaceflight capability since the

2011 retirement of the Space Shuttle, it is still considered a great space power because of its legacy of such capability.

Of the various categories of space powers, it is perhaps *great space power* that is understood best because we have decades of historical experience to draw upon. This chapter will examine great power competition in space—namely the more recent activities of China, Russia, and the United States—to highlight potential challenges and opportunities for international cooperation. After gleaning any lessons or worthwhile considerations from history, topics for the practical implementation of space strategy will be discussed. While the ideas presented here will be in line with the general concepts and principles discussed previously, the ideas and topics in this chapter are thought to be especially salient for great space powers.

China

China's meteoric rise as a space power has been striking. With its manned space program starting in 2003, its anti-satellite testing in 2007, 2010, and 2013, and its plan for a large space station by 2020, the achievements illustrate how rapidly China has matured as a space power.[3] In October 2003, China independently launched and recovered its first *taikonaut*—or astronaut—becoming just the third member of an elite spacefaring club with Russia and the United States. In January 2007, China successfully tested a direct ascent anti-satellite weapon in low Earth orbit, thereby again joining Russia and the United States as one of only three states known to have demonstrated this capability.[4] Regrettably, this test created over 3,000 pieces of trackable space debris.[5] Three years later, China reportedly conducted a test with two micro-satellites performing proximity operations and apparently intentionally "bumping" each other, a capability which only a few countries have at present.[6] It is assessed that this test helped improve Chinese anti-satellite systems. In May 2013, China conducted a self-described "high-altitude science mission," which was assessed by the U.S. Defense Department to be a counter-space test designed to reach satellites in geostationary orbit.[7]

Concerning military operations in space, the intent of the Chinese leadership is difficult to discern. It has been suggested, however, that China understands that to conduct war effectively it must be a space power. Joan Johnson-Freese writes that Beijing understands that it cannot control space all the time, nor does it believe it has to control it. It needs only to buy the time it needs to accomplish its goals by interfering with its opponent's capabilities.[8] Similarly, Chinese military doctrine states that a "soft kill" against an opponent's space system can be achieved by interfering with information systems and ground stations, electromagnetic pulses, camouflage, flare, and deception.[9]

Chinese political and military leadership have concluded that information operations and space capabilities are required to fight and win in future conflicts. Recent Chinese efforts to both exploit and deny space and cyber domains are central to the People's Liberation Army (PLA) focus on fighting and winning future "informationized local wars," of which dominance of outer space, cyber space, maritime, and nuclear domains will play a part.[10]

It is generally understood that the capabilities of the PLA are steadily expanding in every facet. A 2017 Report to the U.S. Congress states:

> The PLA is acquiring a range of technologies to improve China's counterspace capabilities. In addition to the research and possible development of directed-energy weapons and satellite jammers, China is also developing anti-satellite capabilities and probably has made progress on the anti-satellite missile system that it tested in July 2014. China is employing more sophisticated satellite operations and probably is testing dual-use technologies in space that could be applied to counterspace missions.[11]

Moreover, China has concluded that space warfare will be an integral part of future wartime operations.[12] Waging "local war under modern, high-tech conditions" would necessitate space capabilities.[13] In Chinese strategy, space is thought to be important for the advantage it confers when collecting, transmitting, and exploiting information, rather than for space's own sake.[14] Chinese analysis concluded that future joint operations will involve multiple services operating together across significant distances. Victory in future conflicts will not only require unfettered access to space for one's own forces but also the denial of the same ability to the adversary.[15] In the 2013 edition of the Chinese journal *The Science of Military Strategy*, space is said to be the "high ground in wars under informationized conditions," while being tied to the struggles in the future battlegrounds of network space and the electromagnetic spectrum.[16]

Chinese space strategy embraces what would be called joint or multi-domain in Western militaries. Dean Cheng writes that the Chinese embrace an "all aspects unified," thereby viewing the land, sea, air, electromagnetic spectrum, and space in a joint fashion, with operations in each domain contributing to, and receiving support from, the other domains.[17] All operations in these domains are considered to be aimed ultimately at predetermined political ends.[18] Based upon PLA analysts, Chinese military space operations are likely to entail five broad "styles" or mission areas: space deterrence, space blockades, space strike operations, space defense operations, and space information support.[19]

China's growing power and space emphasis may become manifest in mostly peaceful and cooperative ways, or its expanding capabilities may lead to increased competition. Peter Hays has noted, "China presents the global community with both the greatest opportunities and the most difficult challenges for space cooperation."[20] Many security analysts believe that if the United States and others can successfully engage China in effective space cooperation and confidence-building activities, this may help reduce the risks associated with increasing competition. Furthermore, state leaders must avoid the mistake of treating China like the Soviet Union or seeing this relationship through the lens of the Cold War.[21] Brad Roberts has suggested that a potentially more useful historical lens is the interwar period between World Wars I and II, because that period was exemplified by a multipolar world of state competition.[22] If Roberts is correct, historical experience of that period teaches that fear, distrust, miscalculation, and ambiguity should be addressed directly to help avoid any future global conflict.

Russia

Russia's space program is but a shadow of the former Soviet Union's program. The history of the Soviet Union's space program is one of pioneering exploration and accomplishing remarkable feats in space. The Soviet space program launched the first satellite into orbit (Sputnik 1), along with the achievements of having the first man (Yuri Gagarin) and first woman (Valentina Tereshkova) in space. Both the United States and Soviets developed anti-satellite technology and sought to advance the military use of space to protect national interests. Yet the Russian space program barely survived the fall of the Soviet Union, and it has been flagging ever since.

The Soviet Union had a long history of anti-satellite programs and military use of space. The Fractional Orbital Bombardment System—during the 1960s and before the signing of the Outer Space Treaty—was a development program to deliver nuclear warheads to Earth from low Earth orbit.[23] The weapon system was a combination of a low-flying missile and nuclear warhead, and it was designed to take off from the Soviet Union and deorbit when attacking. More importantly, it would not fly over the Arctic to reach the United States, but would rather traverse the southern polar areas and reach the United States via the "backdoor."[24] During the 1960s until the 1980s, the Soviet's Cold War program *Istrebitel Sputnikov*—translated as *satellite killer*—was a program focused on maneuvers to rendezvous with targets to execute a "kamikaze-style" takedown of U.S. space systems, if and when needed.[25]

While the current Russian space program comes from this legacy and still has some notable highpoints—including its heavy-lift rocket capability—the Russian space program is generally languishing compared to days gone by. Much of this downturn in its space program is because of an overall economic and budgetary decline. In referring to the Russian space program, John Logsdon has observed, "Their budget is not adequate to maintain a world-class space effort across the board."[26] The troubles in the Russian space program's supply chain are thought to be symptomatic of their broader economic and financial woes. Logsdon further notes that the Russian space program is undergoing the same problem as the U.S. space program, with core engineers retiring and the younger engineers being attracted to more lucrative employment at purely commercial companies.[27] It has been reported that Roscosmos—the Russian government aerospace agency—has a shrinking workforce, dwindling funds, and systemic corruption—all of which have left the one-time space superpower in a precarious position.[28] Adding to the dismal situation, the Russian space program has seen 15 rocket failures within recent years.[29]

Despite the high number of rocket failures, the Russian heavy lift capability is presently a highlight of Roscosmos. This capability is a vestige of Soviet intercontinental ballistic missile technology and intellectual knowhow. The Russian engines are highly capable, and U.S. companies have used them to power some of their launch vehicles. While using Russian rocket engine technology by U.S. companies may be seen as an unusual arrangement—causing consternation among some members of the U.S. Congress—the performance characteristics of the Russian engines are needed until such time as new indigenous U.S. systems, like the National Aeronautics and Space Administration (NASA) Space Launch System or commercial providers' heavy lift rockets are realized. Under current U.S. legislation, American companies have permission to use the high-performance Russian engines until at least 2022.[30] Moreover, *Soyuz* rockets are presently the only ones capable of carrying astronauts to the International Space Station, until U.S. commercial space companies receive NASA's human spaceflight certification.

During the last decade, there have been noteworthy initiatives in Russian military thinking about the use of space. It has been reported that Russian officials recommended in 2013 that Russia resume research and development of an airborne anti-satellite missile to "be able to intercept absolutely everything that flies from space."[31] Commenting on Russia's renewed interest in space operations, Steve Lambakis notes, "Russia today is experiencing a counterspace revival. Moscow views space as critically important for deterrence and warfighting, and it intends to increase the number of its operational satellites to 150 by 2025."[32] In November 2014, it was reported that a piece of residual debris from the launch of a Russian telecommunication satellite was observed to be maneuvering independently and made a close approach to the rocket stage that boosted it into orbit.[33] As with the Chinese maneuvering capability discussed, these kinds of rendezvous and proximity operations are state-of-the-art.[34] According to an official statement from the Russian Ministry of Defense, in June 2017 the Russians further tested on-orbit satellite inspection capabilities. This is thought to be the first Russian official confirmation of the spacecraft inspector project within the Russian military satellite program.[35]

Much of present Russian space strategy and policy appears to be driven by Russia's desire to have the ideas of "self-sufficiency" and "superiority" permeate all its military space activities.[36] There is still strong, lingering competitiveness between Russia and the United States, and many recent Russian military initiatives in space likely stem from a desire to contest the United States. It is reported that some Russian senior leaders view countering the U.S. space advantage as a critical component of warfighting. Russians largely draw pride from looking back, rather than looking forward. It is thought that Russian nostalgia for the power and prestige of the past has been expertly co-opted by the government under President Vladimir Putin, and the Russian space program is ideally suited for this effort.[37]

While the Russian space program has challenges, the program has repeatedly defied the dire predictions of those foretelling an imminent end. Today, there are efforts to reform and reorganize the Roscosmos state corporation, but questions still linger about its future viability.[38] Time will tell whether Russian leadership can turn their space program around.

United States

Space has been intertwined with national security in the United States since the beginning of the Space Age. U.S. space systems have a history of being vital to monitoring strategic and military developments, including supporting treaty monitoring and arms control verification. The trend of reliance and growth in incorporating space-enabled technologies is expected to continue into the future. Because of the significant investment and integration of space-based capabilities into the U.S. national security space enterprise, it is often concluded by security experts that "The U.S. is more dependent on space than any other nation."[39] To that point, some U.S. security professionals think that potential adversaries perceive space as a weak link in U.S. warfighting capabilities. The 2017 U.S. national security strategy acknowledged this perception in stating, "Others believe that the ability to attack space assets offers an asymmetric advantage and as a result, are pursuing a range of anti-satellite (ASAT) weapons."[40] Consequently, some security analysts may see offensive actions in space as a means of exploiting a perceived U.S. weakness to achieve strategic or military advantage.

U.S. space policy has been mostly consistent during the past several presidential administrations. This includes the policy of acknowledging that all states have the right to explore and use space for peaceful purposes and that space should be used for the benefit of all humanity, in accordance with international law.[41] The U.S. policy has stated repeatedly that the "peaceful purposes" within the 1967 Outer Space Treaty means non-aggressive and allows for space to be used for national and homeland security activities.[42] Admittedly, some countries disagree with this interpretation of "peaceful purposes" language, believing that it precludes any military activities in space.[43] The U.S. interpretation means that space-based systems may lawfully perform essential functions that facilitate military activities on land, in the air, and on and under the sea.[44] Because of its interpretation of "peaceful purposes" within the Outer Space Treaty, U.S. policy has not sought directly the development of new legal regimes, believing the present treaty does not hinder the protection of national interests in space.

Furthermore, U.S. policy is to prevent and deter aggression against space critical infrastructure supporting U.S. national security, while being prepared to defeat attacks and to operate in a degraded environment.[45] This policy includes retaining the right to respond in self-defense, should deterrence fail. The 2017 U.S. National Security Strategy explicitly notes:

> The United States considers unfettered access to and freedom to operate in space to be a vital interest. Any harmful interference with or an attack upon critical components of our space architecture that directly affects this vital U.S. interest will be met with a deliberate response at a time, place, manner, and domain of our choosing.[46]

Overall, U.S. policy has said that the use of force should be in a manner that is consistent with longstanding principles of international law, treaties to which the United States is a party, and the inherent right of self-defense. It should be noted, however, that this legal understanding and recognition of the inherent right of self-defense is from a Western perspective and may not necessarily be shared by either China or Russia.

Non-military measures

> The greatest victory is that which requires no battle.
>
> Sun Tzu[47]

A space power's strategy should be flexible when dealing with the full spectrum of potential conflicts—from large to small in scale. Because future conflict cannot be predicted reliably, a great space power will need to plan for a broad range of both non-military and military activities.

When states decide to act to protect national interests and achieve political ends, they typically will use one or more of the instruments of national power. *Diplomatic*, *information*, *military*, and *economic* are useful categories when considering how states will seek to protect national security interests in space.[48] The diplomatic instrument refers to political efforts between states in the realm of international affairs. This includes initiatives related to international treaties, mutual defense treaties, United Nations Security Council resolutions, demarches, communiques, and international fora. The information instrument refers to facts, data, or instructions in any medium or form, along with its transfer and assigned meaning.[49] The information instrument can include activities associated with traditional broadcast and print news outlets, social media, advertising, Internet information, or the entertainment industry. For the purpose of space strategy, cyber operations and activities will be considered part of the information instrument, even though they may be thought of as more a military function, at times. The military element of national power is the influence and advantage achieved through the application of military presence, coercion, or force. The economic instrument refers to the influence and power that can be gained or lost through trade, commerce, and financial activities.[50] While the four instruments of national power represent means by which the strategy and principles of space warfare can affect other states or organizations, they also represent instruments by which each country or organization can be affected. It is worth emphasizing that the instruments of national power are more often interrelated than not. For example, having significant economic and military power will enable better diplomatic efficacy.

The next discussions will focus on the non-military instruments of power. In those cases when national objectives can be achieved through non-military means—such as diplomatic, economic, and informational efforts—this is effectively winning without fighting, as was lauded by Sun Tzu in his quote at the beginning of this section. Because we have decades of historical experience with diplomacy shaping perceptions and influencing others in space, the diplomatic instrument will be covered briefly during this discussion of great space powers, but will be examined more in Chapters 6 and 7. This is not to imply that diplomacy is unimportant: it is important. Examples of its relevance include the negotiation of arms control treaties based in part on verification using satellites, the drafting and signing of the 1967 Outer Space Treaty, the Apollo-Soyuz Test Project in 1975, and the actions of numerous international organizations that oversee and regulate space activities.

Dividing alliances

Next best is to disrupt his alliances.

Sun Tzu[51]

Great powers can use diplomatic measures to not only build up, but also to tear down. This is the context of Sun Tzu's quote above. While superior space powers may seek to improve their influence through collective and cooperating international activities, they can also decide to reduce the influence and effectiveness of a potential adversary's alliances. When considering attacking an enemy's plans and strategy, Sun Tzu advises to pursue actions that sever or dissolve the enemy's alliances.[52] Doing so may weaken the adversary's overall position. Acts that counter and negate an enemy's coalition and alliances follow under Sun Tzu's general guidance "When he is united, divide him."[53] Dividing alliances are suitable methods to lessen the relative preeminence of a superior power and potentially allowing the conflict to be protracted until the situation changes more in one's favor.

All states, regardless of level of capability, may seek to use diplomacy to contest the power and influence of another, but those considered as having the greatest capability will likely have the means to best advance diplomatic initiatives. Long-term diplomatic initiatives and activities are commensurate with a cumulative strategy, as described by J.C. Wylie. Wylie saw strategy as being of two different kinds: sequential and cumulative. A sequential strategy consists of a series of discrete actions—each which depend upon the one that precede it—that build up all the way to a final decision.[54] He saw all noteworthy land and sea campaigns of the past as reflecting sequential strategies. In contrast, a cumulative strategy does not require ultimate success to be achieved through the results of a sequence of individual actions. Rather, the decision is determined by the sum-total of these actions. Wylie writes, comparing cumulative strategy to the sequential: "The other is the cumulative, the less perceptible minute accumulation of little items piling on top of the other until at some unknown point the mass of accumulated actions may be large enough to be critical."[55] While a sequential strategy is vulnerable if a single effort is thwarted along the way, with a cumulative strategy, individual actions can fail without the entire plan failing.

Consequently, a space super power's effort to divide an adversary's alliances, in line with Sun Tzu's advice, may require many actions to be piled upon each other to achieve the desired result. Such an approach, while being effective, is unlikely be the sole determinant in deciding the outcome between competing states.

Economic considerations

Great space powers will likely have substantial economic influence regarding the use of space-enabled commerce. This economic influence can be used to apply coercive economic pressure against another state, organization, or group to achieve political ends. For millennia, the history of international relations has included states or groups applying economic measure to affect others.

Scott Pace has detailed a model of space policy conflict using the terms *merchants* and *guardians* to describe the differing objectives of public and private sector interests, including how states have historically sought to protect their economic interests.[56] Along those lines, it is expected that great space powers will seek to protect those benefits or advantages coming from space-related commerce, business, and trade. Conversely, these great powers may seek to affect negatively the same of a potential adversary. Such actions are intended to diminish, in part, an adversary's long-term ability to fight wars.

Ultimately, whether efforts to affect negatively an adversary's space-related commerce and trade will depend on a variety of factors. The British artillery officer Charles Callwell's book, *Military Operations and Maritime Preponderance*, is relevant in this regard.[57] He provides sage advice when correcting those who exaggerated sea power's effect on economic warfare, or as he described it the "process of driving an enemy's mercantile flag off the sea and of blockading the hostile coasts."[58] Callwell cautions that the amount of damage that can be inflicted upon an adversary's maritime trade depends upon the volume of that trade, and consequently, varies according to each country.[59]

By logical extension, the efficacy of a great space power to impact the economic instrument of power of another state will depend on the extent of space-reliant commerce, along with the vulnerability of any actions to economic denial. As Bleddyn Bowen comments, "The character of space commerce changes with the type of actor, technology, and economics in play."[60] As a result, the efficacy of economic actions in space will be determined by the affected space power's dependency on space commerce and the character of the distribution of that commercial activity.[61]

Information and cyber warfare

Space activities affect and are affected by information. Consequently, the application of informational effects is an important element in space strategy. Historical experience has shown that information can be used to apply pressure on another's polity to decide in favor of something contrary to their natural inclination or tendency. This thought is in agreement with Clausewitz's thought of compelling the enemy to do one's will by affecting the people, the military, and the governmental leaders, although Clausewitz's context is on the use of force to do so.[62] These kinds of coercive communications present those viewpoints and information in a context that is favorable and beneficial to those sending the information, and the actual information communicated may range from the factual, to the biased, or to the blatantly false.[63]

Along with the message and associated meaning of information that is transmitted via space-based communications, space-reliant data and information can affect the actual conduct of warfare. M.V. Smith describes this ability to collect, assess, and act on space-derived information faster than the enemy in stating:

> Space assets form a ubiquitous global infrastructure—a communications and information backbone—into which friendly forces stationed or deployed anywhere in the world can plug to receive services that increase situational awareness, improve precision engagements, and expedite command and control.[64]

Smith's comment is consistent to John Boyd's observation, orientation, decision, action (OODA) loop theory, which asserts that success in conflict depends on operating within the enemy's decision cycle.[65] According to Boyd, a military with superior observation and orientation capability can make decisions and conduct operations in a manner to which an adversary cannot effectively match or react.[66] Likewise, it is thought that superior space-derived data and information can allow one to have a faster OODA loop, leading to quicker military decisions and actions.

The next consideration is cyber operations, because space systems frequently are a component of cyberspace. Admittedly, cyber operations may be considered under military operations, but based upon how day-to-day cyber operations are commonly not an instrument of violence in combat, cyber operations are considered in this section on non-military measures. Satellites often serve as nodes in a network, and to some extent, the value of satellites is derived from the information they can collect and disseminate on this network.[67] Because celestial lines of communication (CLOCs) are interrelated with the networks that enable space operations and activities, space and cyber operations overlap frequently in operational methods and desired strategic effect.

The cyber domain, or cyberspace, has been defined by Andrew Krepinevich as the world's computer networks, both open and closed, to include the computers themselves, the transactional networks that send data regarding financial transactions, and the networks comprising control systems that enable machines to interact with one another.[68] Like the space domain, the cyber domain utilizes expansive lines of communication involving a global network, along with hubs of activity at server farms or network physical locations.[69] Cyber activities involve international commerce and finance, social media, information sharing, and more recently military operations. Cyberspace is not the sovereign territory of any one state but incorporates both global commons that can be uncontrolled and widely distributed hardware located within sovereign territory. Additionally, cyber operations directly impact and are interrelated with the land, sea, air, and space domains. The lesson to be learned is that space strategy and operations should be integrated with those of the cyber domain.

According to Dean Cheng, Chinese writings emphasize that space dominance entails not only targeting satellites but also ground facilities, including mission control sites and the data links connecting them.[70] He notes that China's space program is not solely devoted to civilian use but provides the PLA with critical pieces of information, deemed essential for both "local wars under high-tech conditions" and "local wars under informationized conditions."[71] Consequently, in the PLA's view, the struggle for space dominance is "part of the larger struggle for information dominance."[72] Likewise in a RAND report, Kevin Poll-peter, Michael Chase and Eric Heginbotham describe that PLA strategists emphasize the crucial role that space plays in the struggle to gain and maintain information dominance, which is seen as deciding the outcome of future military operations.[73] The authors describe how the establishment of the PLA Strategic Support Force, which was announced in December 2015, was not intended to streamline all of China's space enterprise under one command, but was instead intended to facilitate joint operations by providing operational commands with the information-warfare infrastructure necessary to conduct "informationized local wars."[74]

Offensive measures

> Destruction of the enemy forces is always the superior, more effective means, with which others cannot compete.... The commander who wishes to adopt different means can reasonably do so only if he assumes his opponent to be equally unwilling to resort to major battles.
>
> Carl von Clausewitz[75]

By virtue of their capability and influence relative to others, great space powers will need to incorporate offensive strategy as part of an overarching space strategy. Agreeing with the writings of Clausewitz, Colin Gray notes, "Strategic theory advises that while the defense is the stronger form of warfare, offense is more effective."[76] This means that offensive measures—to include the use of military force and violence—will be needed to achieve political ends and bring a conflict to a conclusion. Clausewitz wrote on the importance of offensive action in saying that, "The destruction of the enemy forces is admittedly the purpose of all engagements."[77] According to British maritime strategist Julian Corbett, the stronger power can use offensive actions to obtain positive results or wrest something from the enemy, while also incorporating the "strength and energy" that comes from initiating attack.[78] Corbett believed offensive operations that seek a decision against the enemy's fleet are critical to achieving command of the sea that is both permanent and general.[79]

Offensive strategy will be a necessary component of a great power's success-ful space strategy. Offensive actions impart the advantages coming from initiative and surprise, and offensive actions may let one dictate the pace of hostilities because an enemy may be only capable of reacting to one's own offensive actions. Offensive operations, if successful, may significantly degrade the military capability of an adversary, which could enable more successful offensive operations in the near term.

Despite the need for offensive strategy in war and warfare, offensive strategy has limits. With respect to sea power, Callwell observed that naval forces only provide indirect benefit to land forces when writing, "The effect of sea-power upon land campaigns is in the main strategical. Its influence over the progress of military operations, however decisive this may be, is often only very indirect."[80] Reaching a similar conclusion, Gray describes this in relation to the use of air power in World War II:

> aerial bombing as a threat to deter, inflict damage and pain to coerce, and paralyze or destroy 'works' to provide strategic advantage. It should not be expected to decide by its own unassisted kinetic effort who will win a conflict.[81]

Moreover, offensive strategy and measures are meant to support political ends, which in most cases will be relative to events on land, because that is where most people live. This holds for offensive measures and actions in space strategy. Gray reasons convincingly that:

> Because people live only on land, and belong to security communities organized politically within territorial domains, military behaviour, no matter what its tactical form, ultimately can only have strategic meaning only for the course of events on land. It follows that seapower, airpower, and now spacepower function strategically as enabling factors. The outcome of a war may be decided by action at sea, in the air, or in space, but the war must be concluded on land and usually with reference to land.[82]

So, even though great powers in space are expected to have superior offensive capability, in the end, offensive actions in space are unlikely to be decisive by themselves.

Coercion, compellence, and "small wars"

A great space power will have certain coercive means available to it. Although coercion occurs frequently short of what is considered general war or open hostilities, it will likely include the implicit or explicit threat of some detrimental action, including the potential use of force. Frequently, the purpose of coercion is to affect the decision calculus of a potential adversary, resulting in acquiescence to some demand to achieve political ends.[83]

The Western view of coercion is similar to—albeit not completely the same as—the Chinese view of compellence.[84] The PLA views some military actions during the phases of *military crisis* or *armed conflict* as being outside of actual war, and these actions may seek an adversary to acquiesce on some contentious matter.[85] This thinking is found in PLA writing, as discussed in Chapter 4. The PLA concept of compellence is thought of holistically, where economic, diplomatic, or military actions seek to directly affect an opponent's interests "to compel him to submit to Beijing's will."[86]

Historical experience demonstrates that when considering war or conflict, often it may involve what is considered *small wars*. One of the most definitive and comprehensive works on this topic was the *Small Wars Manual*. Originally printed in 1940, the *Small Wars Manual* describes counterinsurgency and peacekeeping operations. Prior to World War II, the Manual was used by U.S. Marines in preparing for low-intensity conflicts, which in the 1930s were referred to as "small wars." Many of the ideas of the French naval officer and strategist Raoul Castex (1878–1968) were used in formulating the Manual. The Manual defines *small wars* as

> operations undertaken under executive authority, wherein military force is combined with diplomatic pressure in the internal or external affairs of another state whose government is unstable, inadequate, or unsatisfactory for the preservation of life and of such interests as are determined by the foreign policy of our Nation.[87]

Small wars vary from simple demonstrative operations to military intervention, short of actual war. According to the Manual, small wars are not limited in their size, their theater of operations, or their cost in property, money, or lives.[88] Furthermore, U.S. history of intervention through small wars is viewed as a legitimate recourse because, under international law, a state may protect—or demand protection for—its citizens and their property regardless of location.[89] The above definition and justification for small wars acknowledges both the military and non-military nature of warfare, with military strategy being subservient to national policy.

The importance of the Manual is the many subjects covered pertaining to topics are unrelated to direct force-on-force action. These subjects include the roles of the State Department, national government officials, election supervisors, and local resistance fighters. Differing from major military operations between belligerents, small wars are conducted concurrently with a vigorous diplomatic effort to expeditiously reach an agreeable end state.[90] Consequently, the Manual details some of the nuances of how non-military actions are critical in achieving national objectives. The *Small Wars Manual* embraces the thought that mission accomplishment is not predicated on a military victory but by an enduring peace, as exemplified in the statement, "[T]he mission should be accomplished with a minimum loss of life and property and by methods that leave no aftermath of bitterness or render the return to peace unnecessarily difficult."[91]

Because of the prevalence of small wars throughout U.S. history, it could be presumed that future conflicts involving space may be similar to the intent as those described in the *Small Wars Manual*. State leaders may decide to intervene under what is considered legitimate recourse, to protect their citizen's property or interests in space. In the future, space operations may possibly involve measures and efforts to use military force in combination with diplomatic pressure to expeditiously reach an agreeable, lasting, and peaceful solution.

In concluding this topic, a word of caution is warranted. While it should be noted that coercion is at times a suitable strategy, it can go very wrong. If the coercive action causes one on the receiving end of coercion to reassess the situation due to fear, honor, or interest, the result may lead to the unintended escalation of hostilities. Therefore, great care should be given before pursuing a strategy including coercion, especially given the cultural and societal variances in thinking about war, compellence, and deterrence.

Blocking CLOCs

A great space power will want to incorporate measures meant to block an adversary's access to and use of CLOCs.[92] As presented here, this concept is incorporated in the terminology *blocking CLOCs*, as examined in Chapter 2. Using the language in U.S. joint military doctrine, this idea is most analogous to *space negation*, which is defined as "Active defensive and offensive measures to deceive, disrupt, degrade, deny, or destroy and adversary's space capabilities."[93] These measures may include actions against ground, data link, user, and space segments.[94] Because *close blocking* refers to preventing the deployment, launch, or movement of space systems near hubs of activity, it is more defensive in intent because it prevents the adversary from acting or doing something. In contrast, *distant blocking* refers to disrupting, degrading, or denying an adversary's space capabilities far away from the hubs of distribution, but still along distant CLOCs. Distant blocking may result in taking or acquiring the enemy's lines of communication—if they are not already shared between belligerents—while denying the enemy's future use of them. As a result, distant blocking has elements of the offense, and may be especially suitable for great space powers. When faced with blocking along distant CLOCs, the enemy may decide to fight to release itself. Consequently, blocking is a strategy that may be used to force action on terms favorable to the great power implementing the blocking.

Comparable to this idea of blocking CLOCs, the Chinese military writings also discuss "space blockade operations" (*kongjian fengsuo zuozhan*).[95] According to the analysis by Dean Cheng, such operations are intended to intimidate or coerce an adversary.[96] *Space blockades*, according to PLA writings, involve the use of both space and terrestrial forces to prevent an opponent from entering space and from gathering or transmitting information through space. Of note, Chinese writing emphasizes that space blockades have high requirements for precise control, detailed space situational awareness, and highly focused, limited deployment of weapons.[97] According to these writings, there are several kinds of space blockading activities. The first kind is to "blockade terrestrial space facilities" (*hangtian jidi fengsuo*), including launch sites; tracking, telemetry, and control (TT&C) sites; and mission control centers.[98] This seems similar to what was described previously as the maritime-inspired *close blocking*. The blockading activities may use kinetic means, like special operations forces or missiles, or use non-kinetic means, like computer and information network interference. The second method is to "blockade orbits" (*guidao fengsuo*), which can include actually destroying satellites that are in orbit, creating clouds of space debris, or deploying space mines, thereby denying an adversary the easy use of a given orbital track. This thinking is similar to the intent of what was described previously as *distant blocking*. Chinese analysis recognizes the associated risk resulting from damaging third-party space systems, which in turn could lead to unintended consequences. Third, space blockade can include delaying launch windows, so that a satellite may not reach its proper orbit at the correct time.[99] The fourth method is to impose an "information blockade" (*xinxi fengsuo*) by disrupting an opponent's data links between terrestrial control stations and the satellite, thereby interfering with the satellite's control systems or preventing ground control from issuing instructions. This method is thought to achieve a "mission kill."[100]

When providing a caution on naval blockades, Callwell has an observation that is equally relevant to blocking or blockades in space. From looking at maritime history, he concludes that:

> The blockade of the coasts of any state included therein may be an inconvenience to the state, —it may constitute a menace to its prosperity and a check to its advancement, —but such blockade will hardly suffice by itself to coerce that state into sacrificing what it believes to be its rights, or to drive a self-respecting people into purchasing peace by appreciable concessions.[101]

Therefore, while blocking an adversary's CLOCs—whether close or distant— may be beneficial in achieving desired political ends and military advantage, it is unlikely to achieve the victory alone against a determined adversary.

ASATs and Hukkers

Great space powers will need to integrate measures that help ensure one's access to and use of space, while denying the same to potential adversaries. This thinking is nothing new. The intended purpose of anti-satellites weapons (ASATs) is exactly that. During the ASAT testing of the Soviet Union and the United States of the 1970s and 1980s, both countries sought methods of engaging an adversary's satellite in low Earth orbit.[102] Additionally, the self-described science missions of the Chinese may have had the dual-use purpose of demonstrating ASAT capabilities meant to deny others' actions in space. ASAT weapons systems are typically thought of as being launched terrestrially, whether by aircraft, ship, mobile launch vehicle, or fixed ground facility.

By logical extension, there is a similar need to potentially deny an adversary's access to and use of space from orbit. Again, many have written on such a subject, albeit using different terminology. RAND crafted the concept of a "satellite missile" in 1946.[103] Stuart Eves calls the systems *demons*; for Brian Chow it is *stalkers*.[104] Regardless of what these on-orbit or cislunar systems are called, they will need to detect, track, identify, and potentially engage an adversary's satellite or other space system, using the parlance of U.S. joint service doctrine.[105]

Not intending to merely add to the list of candidate terminology or analogies, it is possible to draw upon the history of anti-submarine warfare during the Cold War, to provide a useful mental construct when thinking about offensive operations in space. Indeed, there are operational parallels between anti-submarine warfare and what is required in space. Similar to the functions of space situational awareness to detect, track, identify, and engage, units conducting the anti-submarine mission achieve maritime domain awareness through the phases of search, detect, classify, localize, and when permitted, attack.[106] This process seems analogous to the mission of finding and potentially negating an adversary's satellite from space.

During the 1950–1970s, the U.S. Navy employed hunter-killers task groups— Task Groups Alpha and Bravo—which utilized a small aircraft carrier, fixed-wing aircraft, helicopters, destroyers, and submarines to hunt for and potentially engage enemy submarines.[107] Therefore, the hunter-killer group operated in a multi-domain fashion of sub-surface, surface, and air. In referring to the anti-submarine aircraft, some of these hunter-killer aircraft squadrons were abbreviated and simply referred to as *hukkers*.[108]

Through operational analogy, which admittedly has its limitations, it can be deduced that finding and potentially engaging an adversary's satellite or space systems requires multi-domain solutions.[109] Consequently, land, sea, air, space, and cyber assets may all contribute towards this important mission. To address the threat of an adversary's systems in orbit, hukkers will be needed, while operating in coordination with other assets and, at times, in autonomy to support the hunter-killer mission. Because CLOCs are widely spread, some hukkers should be designed to change orbits or move locations to find and engage another's space asset. If used as a coordinated constellation of satellites, a group of cooperative hukkers could potentially impact an adversary's access and use of CLOCs within a wider region or within an entire orbital regime.

Defensive and offensive strategies integrated

> The defensive form of war is not a simple shield, but a shield made up of well-directed blows.
>
> Carl von Clausewitz[110]

Even though offensive strategy will be a suitable option for great space powers, defensive measures and approaches must not be an afterthought. The defense is the stronger form of war, even in space. The defense is not a strategy that merely "awaits the first blow," but is a strategy of alert expectation, where forces and assets are developed and fielded that protect national interests in space and thoughtfully consider potential threats. During peacetime, space powers should make the necessary preparations to take full advantage of the benefits coming from the defense. This may include mission assurance and resiliency measures discussed previously. Making effective preparations is essential to mitigate the risks and uncertainty associated with great power competition in space.

While agreeing that an offensive strategy during military operations is necessary to achieve victory, defensive strategies in some locations allow for more effective and successful offensive measures in other locations. This would be case when employing the principal of dispersal and concentration. Albeit concentration of assets or effects may be needed, it is still essential to protect and defend those lines of communication that enable offensive efforts. This agrees with Corbett's advice regarding victory at sea. Because sea lines of communication between belligerents are often shared, Corbett declared, "We cannot attack those of the enemy without defending our own."[111] A great power will need to protect and defend its access to those CLOCs that enable offensive operations and effects, and consequently, defensive strategies are frequently appropriate along lines of communication that support offensive action elsewhere.

Chinese space strategy recognizes the interdependence and interplay between offense and defensive measures. The Chinese view space superiority as being achieved with both offensive and defensive counter space operations, while also using multi-domain platforms.[112] China's space strategy writings highlight the need to consider offensive and defensive operations simultaneously to better ensure that offensive and defensive approaches are coordinated, non-interfering, and efficient in their execution.[113]

"A barrier to action" and escalation control

Of emerging, medium, or great powers, only the great space power is likely to be able to command space to the degree that allows making space a barrier against a potential adversary.[114] By definition, great space powers will have a superior ability to potentially establish command of space, as addressed in Chapter 2. Having such superior ability allows for a level of command that is more general and persistent in nature. In contrast, for an adversary who does not have the same level of command in space, these same lines of communication are more likely to become an obstacle to such access and use.

Similar in strategic intent, Corbett wrote of establishing a barrier to action at sea. Corbett observed, "he that commands the sea is at great liberty and may take as much or as little war as he will...."[115] Corbett's description was meant to highlight how the highest capability at sea could enable limited war or prevent the enemy from action. Corbett also referred to the concept using the phrases "the power of isolation" and an "insuperable physical obstacle."[116] By extension, having a superior capability in space—one that includes the ability to command space, thereby ensuring access to and use of CLOCs—a great space power will be better able to defend against enemy action or control undesired escalation by an adversary. Making space a barrier to action is commensurate with defensive strategy, because it seeks to prevent the enemy from doing or taking something.

The concept of a "barrier to action" is readily understood in the other domains of military conflict. In land warfare, defensive barriers include earthworks like trenches or defensive walls like the Great Wall of China and Hadrian's Wall; in maritime warfare, a barrier includes the naval blockade or mining harbors; and in air warfare, a defensive barrier includes integrated air defenses or air defense engagement zones. Even in cyber warfare, defensive measures includes firewalls and "air gaps" in network designs. Space, too, will have orbits and regions where a great power can seek to limit or prevent an adversary's unfettered access to and use of space.

Creating a "barrier to action" will better enable initiation of a conflict having limited political objectives without fearing the enemy's ability to escalate the conflict into an unlimited war in space. Space conflicts with limited aims may include those attempting only to acquire certain CLOCs or those to ensure continued access within a region of space. In contrast, unlimited aims would include the unconditional capitulation of the enemy's populace, military, and leadership.[117] If the strategy is executed adequately, a great space power will have the ability to predict and allocate those resources needed to achieve strategic objectives because the conflict is less likely to spiral out of control into a war of unlimited aims and means.

Space strategy should facilitate the practical execution of war and warfare. The idea of a "barrier in space" may seem a bit far-fetched to be of practical utility. Yet historical examples of the concept are not hard to find. This includes the Brilliant Peebles concept of the late 1980s. Brilliant Pebbles was a U.S. on-orbit initiative—which was never fielded—meant to engage and destroy enemy missiles with non-nuclear weapons as part of a ballistic missile defense system, to keep the U.S. homeland safe.[118] Another potential implementation of a barrier for escalation control may include a mega-constellation of satellites operating in low Earth orbit, to restrict movement of satellites or spacecraft through the orbital regime using non-kinetic methods. The technical requirements for similar approaches would appear feasible given current state-of-the-art space capabilities.

The enemy gets a vote

Even for great space powers, it should be remembered that the enemy gets a vote. Potential adversaries may take actions ahead of time where one's superior space capability is of little consequence to them. This action can include making a conscious effort ahead of time to limit space vulnerabilities or minimize risks in space. This may comprise the thorough incorporation of mission assurance measures into space-based capabilities. This approach may also include a state's leadership deciding to "give up on space" as a domain for protecting national security interests. In cases like this, a great space power may have little to no means of affecting these "space-minimalists."

The writings of Callwell exemplify this thinking that great powers need to consider the strategy of others. Drawing upon his military experience as a British artillery officer, he refers to the writing of Mahan within the context of Britain's unrivaled maritime supremacy; yet, he warns his British counterparts,

> We, with our vast naval resources and noble traditions at sea, are inclined to regard the art of maritime war solely from the point of view of the stronger side. We are prone to forget that when in any set of operations the conditions dictate the adoption of an aggressive attitude to one belligerent, those conditions may dictate the adoption of a Fabian policy to the other belligerent. It is too often forgotten that the destruction of a hostile navy cannot easily be accomplished, even when that navy represents only a relatively speaking feeble fighting force, unless it accepts battle in the open sea.[119]

Therefore, just because a country has extraordinary military capability, the position and action of "the weaker side cannot possibly be ignored."[120]

Additionally, Callwell further elaborates that the strategic environment and geopolitics will shape the nature of any great power competition. In this passage, he compares situations of the United Kingdom, an island-nation that would be devastated if it lost maritime supremacy, and Austria–Hungary, a continental power mostly self-supporting territorially. In referring to results coming from British naval actions and sea power, Callwell writes:

> They can ruin an enemy's maritime commerce. They can blockade the sea board of the opposing belligerent. But their capacity for damaging the foe stops with the shore, —it is limited to the effect which may be caused upon the hostile community by cutting off the sources of supply from oversea. These sources of supply may be vital to the existence of the people; they may be of, comparatively speaking, no importance. A country like the United Kingdom, to which its oversea trade is its life's blood, can be brought to its knees at once by action of a stronger navy. A country like Austria-Hungary, which is virtually self-supporting, which is begirt by productive territory, and which possesses only a modest mercantile marine, may be inconvenienced by hostile sea power, but will never be crushed by it alone.[121]

Put simply, sea power is not guaranteed to have strategic effect in every conflict, as shown by its effect against countries surrounded by water and countries that are land-locked.

This observation holds for space power, as well. Pick your terminology— *space power, command of space, space control,* or *space dominance*—none of these implied conditions are guarantees for victory or achieving political ends. Colin Gray has noted this as well. He writes, "Spacepower must always be useful, but its precise roles and actual strategic utility will be distinctive to each class and case of conflict."[122] The lesson for great space powers is that planning should include conditions where space-related activities will contribute significantly to war's conclusion and those situations where it will not.

Conclusion

This chapter discussed those strategic approaches considered most relevant to great space powers. This includes the balanced use of offensive and defensive strategies to achieve political ends. Because the future is unknowable in any detail, great powers must conduct contingency planning against a full range of contingencies—from small skirmishes to large-scale conflicts. When having the highest capability in space relative to others, great space powers may be able to defend against attacks by potential enemies or control the escalation of hostilities in space.

While the focus here is on space strategy, fighting a great power conflict extending into, through, or from space will likely require an all-domain effort using all instruments of national power. This assessment agrees with the Chinese "all aspects unified" philosophy.[123] As a result, great power space strategy should look to employ its land, sea, air, space and cyber forces in concert with each another to better achieve military success in support of political objectives.

Notes

1 Deganit Paikowsky, "The Space Club—Space Policies and Politics" (paper presented at the 60th International Astronautical Congress, Daejeon, Republic of Korea, October 2009). Also referenced by Scott Pace, "A Space Launch without a Space Program," *38 North*, April 11, 2012, http://38north.org/2012/04/space041112
2 Pace, "A Space Launch without a Space Program."
3 Dean Cheng, *Cyber Dragon: Inside China's Information Warfare and Cyber Operations* (Santa Barbara, CA: Praeger Security International, 2017), 159.
4 Ibid., 160.
5 Ibid.
6 Ibid., 161.
7 Maximilian Betmann, "A Counterspace Awakening? (Part 1)," *The Space Review*, May 22, 2017, www.thespacereview.com/article/3247/1
8 Joan Johnson-Freese, *Space as a Strategic Asset* (New York: Columbia University Press, 2007), 222.
9 Ibid.
10 Dean Cheng, "Are We Ready to Meet the Chinese Space Challenge?" *Spacenews*, July 10, 2017, http://spacenews.com/op-ed-are-we-ready-to-meet-the-chinese-space-challenge/
11 Department of Defense, *Annual Report to Congress: Military and Security Developments Involving the People's Republic of China 2017* (2017), 35, www.defense.gov/Portals/1/Documents/pubs/2017_China_Military_Power_Report.PDF
12 Cheng, *Cyber Dragon*, 155.
13 Ibid., 156.
14 Ibid., 159.
15 Ibid., 157.
16 Shou Xiaosong, ed. *The Science of Military Strategy* (Beijing: Military Science Press, 2013).
17 Cheng, *Cyber Dragon*, 162.

18 Ibid.

19 Ibid., 165.

20 Peter L. Hays, *Space and Security: A Reference Handbook* (Santa Barbara, CA: ABC-CLIO, LLC, 2001), 92.

21 Ibid.

22 Brad Roberts, *The Case for U.S. Nuclear Weapons in the 21st Century* (Stanford, CA: Stanford Security Studies, 2015).

23 Paul B. Stares, *The Militarization of Space: U.S. Policy, 1945–1984* (Ithaca, NY: Cornell University Press, 1985), 80.

24 Braxton Eisel, "The FOBS of War," *Air Force Magazine*, June 2005, www.airforce mag.com/MagazineArchive/Pages/2005/June%202005/0605FOBs.aspx

25 Sergei N. Khrushchev, *Nikita Khrushchev and the Creation of a Superpower* (University Park, PA: Penn State Press, 2010), 351–360. As referenced in Michael Nayak, "Deterring Aggressive Space Actions with Cube Satellite Proximity Operations: A New Frontier in Defensive Space Control," *Air and Space Power Journal* vol. 31 no. 4 (Winter 2017), 94.

26 John Logsdon quoted in Emma Grey Ellis, "Russia's Space Program is Blowing Up. So are its Rockets," *Wired*, December 7, 2016, www.wired.com/2016/12/russias-space-program-blowing-rockets/

27 Ellis, "Russia's Space Program is Blowing Up."

28 Ibid.

29 Asif Siddiqi, "Russia's Space Program is Struggling Mightily," *Slate*, March 21, 2017, www.slate.com/articles/technology/future_tense/2017/03/russia_s_space_program_is_in_trouble.html

30 Valerie Insinna, "RD-180 Ban Thrusts Russian Manufacturer into Uncertain Future," *Defense News*, April 11, 2017, www.defensenews.com/digital-show-dailies/space-symposium/2017/04/11/rd-180-ban-thrusts-russian-manufacturer-into-uncertain-future/

31 Ryan Browne, "Russia Wants to Modify Cold War Missiles to Destroy Asteroids," *CNN*, February 19, 2017, www.cnn.com/2016/02/19/politics/russia-icbm-asteroid-killer/index.html

32 Steve Lambakis, "Foreign Space Capabilities: Implications for U.S. National Security" (National Institute for Public Policy, September 2017), 27, www.nipp.org/wp-content/uploads/2017/09/Foreign-Space-Capabilities-pub-2017.pdf

33 Mike Wall, "Is Russian Mystery Object a Space Weapon?" *Space*, November 19, 2014, www.space.com/27806-russia-mystery-object-space-weapon.html

34 Jim Sciutto, Barbara Starr and Ryan Browne, "Sources: Russia Tests Anti-satellite Weapon," *CNN*, December 21, 2016, www.cnn.com/2016/12/21/politics/russia-satellite-weapon-test/index.html

35 "Cosmos-2519 Released a Satellite Inspector," Russianforces.org, last updated August 23, 2017, accessed August 16, 2018, http://russianforces.org/blog/2017/08/cosmos-2519_released_a_satelli.shtml

36 Jana Honkova, "The Russian Federation's Approach to Military Space and Its Military Space Capabilities," George C. Marshall Policy Outlook (The George C. Marshall Institute, November 2013).

37 Matthew Bodner, "60 Years After Sputnik, Russia is Lost in Space," *Spacenews*, October 4, 2017, https://spacenews.com/60-years-after-sputnik-russia-is-lost-in-space/

38 Ibid.

39 *U.S. Commission to Assess United States National Security, Space Management and Organization*, also known as the *Space Commission Report* (January 11, 2001), 18.

40 The White House, *The National Security Strategy of the United States of America* (December 2017), 31, www.whitehouse.gov/wp-content/uploads/2017/12/NSS-Final-12–18–2017–0905.pdf

41 The White House, *National Space Policy of the United States of America* (June 28, 2010), 2, www.nasa.gov/sites/default/files/national_space_policy_6–28–10. pdf

42 A.R. Thomas and James C. Duncan, *International Law Studies, Volume 73: Annotated Supplement to the Commander's Handbook on the Law of Naval Operations* (Newport, RI: Naval War College, 1999), 149–150.

43 Ibid., 149.

44 Thomas and Duncan, *International Law Studies*, 150.

45 Department of Defense and Office of the Director of National Intelligence, *National Security Space Strategy* (January 2011), 11.

46 The White House, *The National Security Strategy of the United States of America* (December 2017), 31.

47 Sun Tzu, *The Art of War* (*c*.400–320 BC), 3.2.

48 "Joint Forces Command Glossary," U.S. Joint Forces Command, accessed September 2, 2004, www.jfcom.mil/about/glossary/html. Diplomatic, Information, Military and Economic (DIME) are areas of national power that are used in "effects-based" operations.

49 "Joint Forces Command Glossary," U.S. Joint Forces Command.

50 Yuan-Li Wu, *Economic Warfare* (New York: Prentice-Hall, 1952). This reference gives an overview on the methods of using economic influence against others.

51 Sun Tzu, *The Art of War*, (*c*.400–320 BC), 3.5.

52 Sun Tzu, *The Art of War*, trans. Samuel B. Griffith (Oxford: Oxford University Press, 1963), 77–78.

53 Ibid., 69.

54 J.C. Wylie, *Military Strategy: A General Theory of Power Control*, with introduction by John B. Hattendorf (New Brunswick, NJ: Rutgers University Press, 1967; reprint, Annapolis, MD: Naval Institute Press, 1989), 22–27.

55 Ibid., 24.

56 Scott Pace, *Merchants and Guardians: Balancing U.S. Interests in Space Commerce* (Santa Monica, CA: RAND Corporation, 1999), www.rand.org/pubs/reprints/RP787. html.

57 Charles E. Callwell, *Military Operations and Maritime Preponderance: Their Relations and Interdependence*, with introduction by Colin S. Gray (London: William Blackwood and Sons, 1905; reprint, Annapolis, MD: Naval Institute Press, 1996), 443.

58 Ibid., 170.

59 Ibid., 176–177.

60 Bleddyn E. Bowen, *Spacepower and Space Warfare: The Continuation of Terran Politics by Other Means* (PhD Thesis, Aberystwyth University, 2015), 210.

61 Ibid.

62 Carl von Clausewitz, *On War*, trans. and ed. Michael Howard and Peter Paret (Princeton, NJ: Princeton University Press, 1989), 75, 89.

63 This concept is also sometimes referred to as *information operations*, *information warfare*, and *psychological warfare*. "Joint Forces Command Glossary," U.S. Joint Forces Command.

64 M.V. Smith, "Spacepower and the Strategist," in *Strategy: Context and Adaption from Archidamus to Airpower*, eds. Richard J. Bailey Jr., James W. Forsyth Jr., and Mark O. Yeisley (Annapolis, MD: Naval Institute Press, 2016), 171.

65 Grant T. Hammond, *The Mind of War: John Boyd and American Security* (Washington: Smithsonian Institution, 2001), 4–5.

66 Ibid.

67 Committee on National Security Space Defense and Protection, "National Security Space Defense and Protection" (The National Academies of Sciences, Engineering, and Medicine, 2016), 13.

68 Andrew F. Krepinevich, "Cyber Warfare: A 'Nuclear Option'?" (Center for Strategic and Budgetary Assessments, 2012), 8, https://csbaonline.org/research/publications/cyber-warfare-a-nuclear-option

69 John J. Klein, "Some Principles of Cyber Strategy," *International Security Network* (August 21, 2014), www.files.ethz.ch/isn/187930/ISN_182955_en.pdf

70 Cheng, *Cyber Dragon*, 176.

71 Ibid., 160.

72 Ibid., 176.

73 Kevin L. Pollpeter, Michael S. Chase, and Eric Heginbotham, *The Creation of the PLA Strategic Support Force and Its Implications for Chinese Military Space Operations* (Santa Monica, CA: RAND, 2017), 1; www.rand.org/content/dam/rand/pubs/research_reports/RR2000/RR2058/RAND_RR2058.pdf

74 Ibid.

75 Carl von Clausewitz, *Vom Kriege*, erster Band (Berlin: Ferdinand Dümmler, 1832), 48–50.

76 Colin S. Gray, *Airpower for Strategic Effect* (Maxwell Air Force Base, AL: Air University Press, 2012), 292.

77 Clausewitz, *On War*, 236.

78 Julian S. Corbett, *Some Principles of Maritime Strategy* (London: Longmans, Green and Co., 1911; reprint, Annapolis, MD: Naval Institute Press, 1988), 34.

79 Ibid., 167.

80 Charles E. Callwell, *The Effect of Maritime Command on Land Campaigns since Waterloo* (Edinburgh: William Blackwood and Sons, 1897), 29.

81 Gray, *Airpower for Strategic Effect*, 297

82 Colin S. Gray, *Modern Strategy* (Oxford: Oxford University Press, 1999), 259.

83 John J. Klein, *Space Warfare: Strategy, Principles and Policy* (Abingdon: Routledge, 2006), 63.

84 Dean Cheng, "Evolving Chinese Thinking about Deterrence: What the United States Must Understand About China and Space," Backgrounder No. 3298 (The Heritage Foundation, March 29, 2018), 2, http://report.heritage.org/bg3298

85 Peng Guangqian and Yao Youzhi, eds., *The Science of Military Strategy*, (Beijing: Military Science Publishing House, 2001), 230. Referenced in Alison A. Kaufman and Daniel M. Hartnett, "Managing Conflict: Examining Recent PLA Writings on Escalation Control," 54.

86 Mark A. Stokes, "The Chinese Joint Aerospace Campaign: Strategy, Doctrine, and Force Modernization," in "China's Revolution in Doctrinal Affairs: Emerging Trends in the Operational Art of the Chinese People's Liberation Army," eds. James Mulvenon and David M. Finkelstein (CNA, 2005), 226–227.

87 United States Marine Corps, *Small Wars Manual* (Washington, DC: 1940; reprint, 1990), section 1–1.

88 Ibid., section 1–1.

89 Ibid., section 1–5. The *Small Wars Manual* divides small wars into five distinct phases. Phase 1 is the initial demonstration or arrival of landing forces; phase 2 is the arrival of reinforcements and general military operations; phase 3 is assuming control of local executive, legislative, and judicial agencies; phase 4 is conducting routine police functions; phase 5 is the withdrawal of forces from the theater of operations.

90 Ibid., section 1–5.

91 Ibid., section 2–3.

92 Klein, *Space Warfare*, 91–92.

93 Joint Chiefs of Staff, *Space Operations*, Joint Publication 3–14 (May 29, 2013), II–8.

94 Ibid.

95 Cheng, "Evolving Chinese Thinking About Deterrence: What the United States Must Understand About China and Space," 4.

96 Ibid., 4–5.

97 Ibid.
98 Ibid.
99 Cheng, *Cyber Dragon*, 167.
100 Ibid.
101 Callwell, *Military Operations*, 173.
102 Steven Lambakis, *On the Edge of Earth: The Future of American Space Power* (Lexington, KY: The University Press of Kentucky, 2001), 121–123.
103 Ibid., 97.
104 Stuart Eves, "Angels and Demons: Cooperative and Non-cooperative Formation Flying with Small Satellites" (presentation, Surrey Satellite Technology Limited, London, 2008), 2; Brian G. Chow, "Stalkers in Space: Defeating the Threat," *Strategic Studies Quarterly* vol. 11 no. 2 (Summer 2017), 100.
105 Erin Salinas, "Space Situational Awareness is Space Battle Management," *Air Force Space Command*, May 16, 2018, www.afspc.af.mil/News/Article-Display/Article/1523196/space-situational-awareness-is-space-battle-management/
106 David P. Finch, "Anti-submarine Warfare (ASW) Capability Transformation: Strategy of Response to Effects Based Warfare" (presented at the 16th International Command and Control Research and Technology Symposium, Quebec City, Canada, June 21–23, 2011), 2, www.dtic.mil/dtic/tr/fulltext/u2/a547026.pdf
107 Robert C. Manek, "Overview of U.S. Navy Antisubmarine Warfare (ASW) Organization during the Cold War Era," NUWC-NPT Technical Report 11,890 (Naval Undersea Warfare Center Division, August 12, 2008), 8–9, www.dtic.mil/dtic/tr/fulltext/u2/a487974.pdf
108 For instance, Air Anti-Submarine Squadron Two-Eight (VS-28) was referred to as "The World Famous Hukkers" in the 1960–1970s. "Squadrons and Wings," Viking Association, accessed August 15, 2018, www.vikingassociation.com/squadrons-and-wings.php#vs28
109 John B. Sheldon, *Reasoning by Strategic Analogy: Classical Strategic Thought and the Foundations of a Theory of a Space Power* (Ph.D. Thesis, University of Reading, 2007).
110 Carl von Clausewitz, *Vom Kriege*, dritter Band (Berlin: Ferdinand Dümmler, 1832), 144.
111 Corbett, *Some Principles of Maritime Strategy*, 100.
112 Shou Xiaosong, ed., *The Science of Military Strategy*, 182. As referenced in Pollpeter, Chase, and Heginbotham, *The Creation of the PLA Strategic Support Force*, 9.
113 Pollpeter, Chase, and Heginbotham, *The Creation of the PLA Strategic Support Force*.
114 But a medium power can make space a "weak" barrier. See Chapter 6.
115 Original paraphrase is from Corbett, *Some Principles of Maritime Strategy*, 58. Attributed to Francis Bacon, with original reference from *Essays 29*, "Of the True Greatness of Kingdoms and Estates" (1597).
116 Corbett, *Some Principles of Maritime Strategy*, 57–59.
117 Clausewitz, *On War*, 89.
118 Simon P. Worden, "Space Control for the 21st Century: A Space 'Navy' Protecting the Commercial Basis of America's Wealth," in *Spacepower for a New Millennium: Space and National Security*, eds. Peter L. Hays, James M. Smith, Alan R. Van Tassel, and Guy M. Walsh (New York: McGraw-Hill, 2000), 235; Stares, *The Militarization of Space*, 225.
119 Callwell, *Military Operations*, 52–53.
120 Ibid., 91–92.
121 Ibid., 170.
122 Gray, *Modern Strategy*, 264.
123 Cheng, *Cyber Dragon*, 162.

Bibliography

Bacon, Francis. "Of the True Greatness of Kingdoms and Estates," in *The Essays*. 1601; reprint, Adelaide: The University of Adelaide, 2014.

Betmann, Maximilian. "A Counterspace Awakening? (Part 1)." *The Space Review*. May 22, 2017. www.thespacereview.com/article/3247/1

Bodner, Matthew. "60 Years After Sputnik, Russia is Lost in Space." *Spacenews*. October 4, 2017. https://spacenews.com/60-years-after-sputnik-russia-is-lost-in-space/

Bowen, Bleddyn E. *Spacepower and Space Warfare: The Continuation of Terran Politics by Other Means*. PhD Thesis, Aberystwyth University, 2015.

Browne, Ryan. "Russia Wants to Modify Cold War Missiles to Destroy Asteroids." *CNN*. February 19, 2017. www.cnn.com/2016/02/19/politics/russia-icbm-asteroid-killer/index.html

Callwell, Charles E. *Military Operations and Maritime Preponderance: Their Relations and Interdependence*. With introduction by Colin S. Gray. London: William Blackwood and Sons, 1905; reprint, Annapolis, MD: Naval Institute Press, 1996.

Callwell, Charles E. *The Effect of Maritime Command on Land Campaigns since Waterloo*. Edinburgh: William Blackwood and Sons, 1897.

Cheng, Dean. "Are We Ready to Meet the Chinese Space Challenge?" *Spacenews*. July 10, 2017. http://spacenews.com/op-ed-are-we-ready-to-meet-the-chinese-space-challenge/

Cheng, Dean. "Evolving Chinese Thinking about Deterrence: What the United States Must Understand about China and Space." Backgrounder No. 3298. The Heritage Foundation, March 29, 2018. http://report.heritage.org/bg3298

Cheng, Dean. *Cyber Dragon: Inside China's Information Warfare and Cyber Operations*. Santa Barbara, CA: Praeger Security International, 2017.

Chow, Brian G. "Stalkers in Space: Defeating the Threat." *Strategic Studies Quarterly* vol. 11 no. 2 (Summer 2017): 82–116.

Clausewitz, Carl von. *On War*. Translated and edited by Michael Howard and Peter Paret. Princeton, NJ: Princeton University Press, 1989.

Clausewitz, Carl von. *Vom Kriege*, dritter Band. Berlin: Ferdinand Dümmler, 1832.

Clausewitz, Carl von. *Vom Kriege*, erster Band. Berlin: Ferdinand Dümmler, 1832.

Committee on National Security Space Defense and Protection. "National Security Space Defense and Protection." The National Academies of Sciences, Engineering, and Medicine, 2016.

Corbett, Julian S. *Some Principles of Maritime Strategy*. London: Longmans, Green and Co., 1911; reprint, Annapolis, MD: Naval Institute Press, 1988.

"Cosmos-2519 Released a Satellite Inspector." Russianforces.org. Last updated August 23, 2017. Accessed August 16, 2018. http://russianforces.org/blog/2017/08/cosmos-2519_released_a_satelli.shtml

Department of Defense and Office of the Director of National Intelligence. *National Security Space Strategy*. January 2011.

Department of Defense. *Annual Report to Congress: Military and Security Developments Involving the People's Republic of China 2017*. 2017. www.defense.gov/Portals/1/Documents/pubs/2017_China_Military_Power_Report.PDF

Eisel, Braxton. "The FOBS of War." *Air Force Magazine*. June 2005. www.airforcemag.com/MagazineArchive/Pages/2005/June%202005/0605FOBs.aspx

Ellis, Emma Grey. "Russia's Space Program is Blowing Up. So are its Rockets." *Wired*. December 7, 2016. www.wired.com/2016/12/russias-space-program-blowing-rockets/

Eves, Stuart. "Angels and Demons: Cooperative and Non-cooperative Formation Flying with Small Satellites." Presentation. Surrey Satellite Technology Limited, London, 2008.

Finch, David P. "Anti-submarine Warfare (ASW) Capability Transformation: Strategy of Response to Effects Based Warfare." Presented at the 16th International Command and Control Research and Technology Symposium, Quebec City, Canada, June 21–23, 2011. www.dtic.mil/dtic/tr/fulltext/u2/a547026.pdf

Gray, Colin S. *Airpower for Strategic Effect*. Maxwell Air Force Base, AL: Air University Press, 2012.

Gray, Colin S. *Modern Strategy*. Oxford: Oxford University Press, 1999.

Hammond, Grant T. *The Mind of War: John Boyd and American Security*. Washington: Smithsonian Institution, 2001.

Hays, Peter L. *Space and Security: A Reference Handbook*. Santa Barbara, CA: ABC-CLIO, LLC, 2011.

Honkova, Jana. "The Russian Federation's Approach to Military Space and Its Military Space Capabilities." George C. Marshall Policy Outlook. The George C. Marshall Institute, November 2013.

Insinna, Valerie. "RD-180 Ban Thrusts Russian Manufacturer into Uncertain Future." *Defense News*. April 11, 2017. www.defensenews.com/digital-show-dailies/space-symposium/2017/04/11/rd-180-ban-thrusts-russian-manufacturer-into-uncertain-future/

Johnson-Freese, Joan. *Space as a Strategic Asset*. New York: Columbia University Press, 2007.

Joint Chiefs of Staff. *Space Operations*. Joint Publication 3–14. May 29, 2013.

"Joint Forces Command Glossary." U.S. Joint Forces Command. Accessed September 2, 2004. www.jfcom.mil/about/glossary/html.

Kaufman, Alison A. and Daniel M. Hartnett. "Managing Conflict: Examining Recent PLA Writings on Escalation Control." CNA, February 2016.

Khrushchev, Sergei N. *Nikita Khrushchev and the Creation of a Superpower*. University Park, PA: Penn State Press, 2010.

Klein, John J. "Some Principles of Cyber Strategy." *International Security Network* (August 21, 2014). www.files.ethz.ch/isn/187930/ISN_182955_en.pdf

Klein, John J. *Space Warfare: Strategy, Principles and Policy*. Abingdon: Routledge, 2006.

Krepinevich, Andrew F. "Cyber Warfare: A 'Nuclear Option'?" Center for Strategic and Budgetary Assessments, 2012. https://csbaonline.org/research/publications/cyber-warfare-a-nuclear-option

Lambakis, Steve. "Foreign Space Capabilities: Implications for U.S. National Security." National Institute for Public Policy, September 2017. www.nipp.org/wp-content/uploads/2017/09/Foreign-Space-Capabilities-pub-2017.pdf

Lambakis, Steve. *On the Edge of Earth: The Future of American Space Power*. Lexington, KY: The University Press of Kentucky, 2001.

Manek, Robert C. "Overview of U.S. Navy Antisubmarine Warfare (ASW) Organization during the Cold War Era." NUWC-NPT Technical Report 11,890. Naval Undersea Warfare Center Division, August 12, 2008. www.dtic.mil/dtic/tr/fulltext/u2/a487974.pdf

Nayak, Michael. "Deterring Aggressive Space Actions with Cube Satellite Proximity Operations: A New Frontier in Defensive Space Control." *Air and Space Power Journal* vol. 31 no. 4 (Winter 2017): 92–102.

Pace, Scott. "A Space Launch without a Space Program." *38 North*. April 11, 2012. http://38north.org/2012/04/space041112

Pace, Scott. *Merchants and Guardians: Balancing U.S. Interests in Space Commerce.* Santa Monica, CA: RAND Corporation, 1999. www.rand.org/pubs/reprints/RP787. html

Paikowsky, Deganit. "The Space Club—Space Policies and Politics," Paper presented at the 60th International Astronautical Congress. Daejeon, Republic of Korea, October 2009).

Peng Guangqian and Yao Youzhi, eds. *The Science of Military Strategy.* Beijing: Military Science Publishing House, 2001.

Pollpeter, Kevin L., Michael S. Chase, and Eric Heginbotham. *The Creation of the PLA Strategic Support Force and Its Implications for Chinese Military Space Operations.* Santa Monica, CA: RAND Corporation, 2017. www.rand.org/content/dam/rand/pubs/research_reports/RR2000/RR2058/RAND_RR2058.pdf

Roberts, Brad. *The Case for U.S. Nuclear Weapons in the 21st Century.* Stanford, CA: Stanford Security Studies, 2015.

Salinas, Erin. "Space Situational Awareness is Space Battle Management." *Air Force Space Command.* May 16, 2018. www.afspc.af.mil/News/Article-Display/Article/1523196/space-situational-awareness-is-space-battle-management/

Sciutto, Jim, Barbara Starr and Ryan Browne. "Sources: Russia Tests Anti-satellite Weapon." *CNN.* December 21, 2016. www.cnn.com/2016/12/21/politics/russia-satellite-weapon-test/index.html

Sheldon, John B. *Reasoning by Strategic Analogy: Classical Strategic Thought and the Foundations of a Theory of a Space Power.* Ph.D. Thesis, University of Reading, 2007.

Shou Xiaosong, ed. *The Science of Military Strategy.* Beijing: Military Science Press, 2013.

Siddiqi, Asif. "Russia's Space Program is Struggling Mightily." *Slate.* March 21, 2017. www.slate.com/articles/technology/future_tense/2017/03/russia_s_space_program_is_in_trouble.html

Smith, M.V. "Spacepower and the Strategist." In *Strategy: Context and Adaption from Archidamus to Airpower,* edited by Richard J. Bailey Jr., James W. Forsyth Jr., and Mark O. Yeisley, 157–185. Annapolis, MD: Naval Institute Press, 2016.

"Squadrons and Wings." Viking Association. Accessed August 15, 2018. www.viking association.com/squadrons-and-wings.php#vs28

Stares, Paul B. *The Militarization of Space: U.S. Policy, 1945–1984.* Ithaca, NY: Cornell University Press, 1985.

Stokes, Mark A. "The Chinese Joint Aerospace Campaign: Strategy, Doctrine, and Force Modernization." In "China's Revolution in Doctrinal Affairs: Emerging Trends in the Operational Art of the Chinese People's Liberation Army," edited by James Mulvenon and David M. Finkelstein, 221–305. CNA, 2005.

Sun Tzu. *The Art of War. c.*400–320 BC.

Sun Tzu. *The Art of War.* Translated by Samuel B. Griffith. Oxford: Oxford University Press, 1963.

The White House. *National Space Policy of the United States of America.* June 28, 2010. www.nasa.gov/sites/default/files/national_space_policy_6-28-10.pdf

The White House. *The National Security Strategy of the United States of America.* December 2017. www.whitehouse.gov/wp-content/uploads/2017/12/NSS-Final-12-18-2017-0905.pdf

Thomas, A.R. and James C. Duncan. *International Law Studies, Volume 73: Annotated Supplement to the Commander's Handbook on the Law of Naval Operations.* Newport, RI: Naval War College, 1999.

U.S. Commission to Assess United States National Security, Space Management and Organization, also known as the *Space Commission Report*. January 11, 2001.

United States Marine Corps. *Small Wars Manual*. Washington, DC: 1940; reprint, 1990.

Wall, Mike. "Is Russian Mystery Object a Space Weapon?" *Space*. November 19, 2014. www.space.com/27806-russia-mystery-object-space-weapon.html

Worden, Simon P. "Space Control for the 21st Century: A Space 'Navy' Protecting the Commercial Basis of America's Wealth." In *Spacepower for a New Millennium: Space and National Security*, edited by Peter L. Hays, James M. Smith, Alan R. Van Tassel, and Guy M. Walsh, 225–237. New York: McGraw-Hill, 2000.

Wylie, J.C. *Military Strategy: A General Theory of Power Control*. With introduction by John B. Hattendorf. New Brunswick, NJ: Rutgers University Press, 1967; reprint, Annapolis, MD: Naval Institute Press, 1989.

Yuan-Li Wu. *Economic Warfare*. New York: Prentice-Hall, 1952.

6 Space strategy for medium powers

The focus of this chapter is on strategy for medium space powers. As with great space powers, the fundamental purpose of any medium space power's space strategy should be to ensure access to and use of celestial lines of communication to support national objectives, whether during peace or conflict. When compared to the strategies of great powers, however, the strategies of medium powers are often different due to a medium power's desire to act independently, while being comparatively more constrained by available material and fiscal resources than most great powers. This chapter addresses the space programs of India and Iran, and then it examines non-military measures, offensive and defensive strategies, along with the topics of "force in being," limited war, dispersal and concentration, and guardians and resiliency. Even though the focus of this chapter is the preferred strategies for medium space powers, many of the ideas can be applicable to both great and emerging space powers as well, depending on the situation.

Discussions about the space strategy for medium powers are possibly more relevant than those of great power strategy because there are more medium powers than great powers. As described in the previous chapter, Deganit Paikowsky categorized medium space powers as those states with the indigenous capability to launch, develop, and control satellites, while being minus any indigenous human spaceflight capability.[1] Therefore, medium space powers include the European Space Agency (ESA), the European member states of ESA that support ESA's space launch capabilities, Japan, India, Israel, Ukraine (which inherited its launch capability through the former Soviet Union), and Iran.[2] Admittedly, the number of medium space powers is limited due to the definitional inclusion of needing an indigenous satellite launch capability, and the definition does not take into account how the space powers are trending currently relative to others. This condition may cause some space analysts to take exception with the definition's utility. Nevertheless, the definition allows for medium space powers to be differentiated from those of great or emerging space powers, using a common metric for comparison.

When thinking about medium powers—along with their interests and strategies—maritime strategy can provide a suitable framework for discussion. This is especially the case within the writings of Charles Callwell, Raoul Castex, and J.R. Hill.

In his book *Maritime Strategy for Medium Powers* (1986), J.R. Hill describes, "Medium powers then lie between the self-sufficient and the insufficient."[3] For this reason, the space strategies of medium space powers are expected to be different from either emerging or great space powers. Hill states that a medium power's fundamental security objective is *"to create and keep under national control enough means of power to initiate and sustain coercive actions whose outcome will be the preservation of its vital interests."*[4] In an article written almost 15 years later than his book, Hill further elaborates:

> The medium power, by its very nature, is likely to have few resources to spare for the exercise of power beyond what is necessary to safeguard and, where possible, further its vital interests of territorial integrity, political independence and betterment. The extent of those vital interests needs to be carefully assessed. But once that has been done, then the medium power will want to keep the levers of power in its own hands to the maximum extent possible.[5]

Hill understands that for a medium power, there is one constant, nagging question: "[H]ow much should one be able to do on one's own before being forced to call on help from an ally—whether that ally be formal or informal, superpower or another medium power?"[6] Not intending to be glib, he answers "as much as the situation demands."[7] A medium power must be a primary contributor in the actions required, even if other states or organizations are sooner or later engaged in its support.[8] This consideration is typically at the forefront of a medium power's strategy development.

Within maritime history and strategy, medium powers have been often considered wrongly under the lens of the *haves* and *have nots*. This bipolar approach has led to the tendency of concluding incorrectly that if a "power could not do everything in a war, it could do nothing."[9] Maritime historian and strategist Geoffrey Till has made a similar observation. Referring to thinking of maritime strategy and great power competition, Till notes that *command of the sea* has routinely been mistaken to mean commanding all or none of the sea at all times, rather than some level of control in limited areas and for limited times.[10] He insists that "the concept is relevant to small navies as it is to big ones."[11]

This is an applicable lesson for space strategy development. Although medium space powers frequently will not have the same options as great space powers for protecting or promoting their objectives in space, medium space powers can still seek to protect their space-related vital interests using their available means to the fullest extent possible. Because of their more limited resources and available means, medium space powers are more likely to carefully husband their resources, which may lead to more carefully considering strategies and force planning, when compared to great powers. Also, medium space powers maybe be able to bring specialized or niche capabilities into cooperative relationships and alliances with other space powers. If this capability is sufficiently specialized and advantageous, great powers can become reliant upon the medium power's technologies, capabilities, or processes.

In the next two sections, the activities of two medium space powers—India and Iran—will be described, to place these powers and strategies in perspective to other countries. These two powers were chosen because of their contrasting reasons for their space activities, along with the disparate range in their space capabilities. Then, some areas of space strategy most relevant to medium space powers will be described.

India

Historically, the Indian government has taken the view that the use of outer space should be primarily for civilian benefit and the development of a national space infrastructure can be broadly advantageous to its citizens. This view was indeed the belief of Vikram Sarabhai, considered to be the founding father of the Indian space program.[12] According to the Indian Space Research Organisation (ISRO), the current vision of India's space program is to "Harness space technology for national development, while pursuing space science research and planetary exploration."[13] This approach has resulted in a capable Indian space program that has indigenously developed launch, satellite, and ground systems for the civilian benefit. Some practical applications include resource monitoring, meteorology, and disaster management.[14] India has a long history with launch and satellites. India launched its first sounding rocket—a U.S. supplied Nike-Apache—in November 1963.[15] In July 1980, India became only the seventh country to have an indigenous satellite launch capability with the launch of its Rohini RS-1 satellite.[16]

India's military use of space is illustrative of how space assets can support joint and dispersed operations. In May 1999, India launched its first ocean observation satellite, Oceansat. The satellite monitors the India Ocean and Bay of Bengal, and the Indian navy benefits greatly from this maritime observation capability.[17] Additionally, both the Indian army and air force routinely operate across the Indian subcontinent, and both services benefit from satellite-based communications networks that facilitate the command and control of their forces. In June 2010, India established an Integrated Space Cell, located in the Integrated Defense Headquarters.[18] The Integrated Space Cell is in charge of defense-related space capability requirements and is comprised of all the services of the armed forces, the Department of Space, and ISRO.[19] A senior ISRO official concluded that because India has invested a great deal into its space capabilities, "it becomes necessary to protect them from adversaries. There is a need to look at means of securing these."[20] Consequently, ensuring India's ability to use its space capabilities is considered a vital national interest.

India is still making great strides in its satellite and launch capabilities. Between 1999 and 2016, India launched over 50 foreign satellites into orbit.[21] In February 2017, the ISRO launched 104 satellites on a single Polar Satellite Launch Vehicle (PSLV) rocket, setting a world record.[22] In January 2018, India launched a Cartosat-2 Earth observation satellite, along with 30 other micro- and nanosatellites from six different countries.[23] According to the ISRO, Cartosat-2 can perform a wide variety of cartographic applications including: coastal land use and regulation, utility management like road network monitoring, water distribution, creation of land use maps, and change detection to bring out geographical and man-made features.[24] The Indian press has reported that Cartosat-2 also has dual-use national security applications, especially in monitoring activity along India's borders.

Additional Indian achievements include scientific missions to the Moon and Mars. India sent a spacecraft, Chandrayaan-1, to orbit the Moon in 2008.[25] Another mission to the Moon is planned for 2019 using the Chandrayaan-2 spacecraft. This mission aims to demonstrate that India can land a spacecraft and drive a rover on the Moon.[26] In 2013, the ISRO sent an exploratory probe, Mangalyaan, to Mars.[27] By successfully placing the Mangalyaan spacecraft into orbit around Mars, India achieved a spacecraft accomplishment that only four other countries had achieved previously.[28] According to Joan Johnson-Freese, the Mangalyaan mission was driven to a large extent for prestige reasons, especially with respect to China.[29]

Recent changes in the geopolitical environment are putting additional pressures on India's space program. Johnson-Freese notes that the scope of India's civil and military space programs has broadened considerably in the recent decade, driven mostly by geo-strategic reasons.[30] As more countries have incorporated space into their security capabilities, this has become a more attractive approach to India as well. For example, after the Chinese 2007 anti-satellite test, it is reported that Indian officials began to consider whether India should have its own anti-satellite capability and compete regionally with China.[31]

Given its growing trend in space launch and satellite capabilities, it is expected that India will continue to be a notable space power, both regionally and globally. The country will likely continue to have a significant international role in space launch, while receiving economic return on its investment from other countries. Finally, it is anticipated Indian space activities will increase in national security areas, as more dual-use space technologies are incorporated into its defense and military operations.

Iran

Iran's space program has a history of irregular and inconsistent progress. Iran is one of the 24 founding members of the United Nations Committee on the Peaceful Uses of Outer Space established in 1958, and United Nations Office of Outer Space Affairs' records show that Iran has been involved in international space dialogue since the earliest days of the office's founding.[32] In the late 1970s, Iran sought to establish a satellite communications program; yet, it could not achieve the objective indigenously and consequently turned to the Soviet Union, China, North Korea, and India for help.[33] The Iranian Revolution (1979) and the war with Iraq (1980–1988) derailed many space program initiatives of this period. The country established it space agency in 2004, added significant investment in 2010, and continues to be in a seemingly perpetually nascent condition of launching rockets and satellites.[34]

Some security experts believe Iran's space program was driven initially by prestige reasons.[35] Evidence supporting this view can be found in Iranian President Mahmoud Ahmadinejad saying, "When we launch a satellite into space, there is a huge boost in the morale of the public."[36] While prestige may be a factor in its space ambitions, there are military implications because of the dual-use applications of rocket launch for ballistic missile technology. Much of Iran's technological knowhow is a direct result of its past work on short- and medium-range ballistic missiles, and the country's interest in ballistic missile technology can be traced to the war with Iraq in the 1980s.[37]

Iran possesses a proven space launch vehicle—the Safir rocket. The space launch capability and technical knowhow that Iran possess are largely based upon North Korean missile technology.[38] Presently, Iran is developing a more capable launch vehicle known as the Simorgh, but it has experienced delays. Both the Safir and Simorgh launch vehicles are liquid-fueled rockets and are said to be launched from a single space launch facility after a significant set-up period.[39] In April 2016, the first known test of the Simorgh was reported as a "partial success" that did not reach orbit.[40] During a second test launch in July 2017, Iranian media reported the launch event as successful, but other reporting described the test as a catastrophic failure because no objects reached the intended orbit.[41] When considering its dual-use capability, a Simorgh-type space launch vehicles is estimated to have a 7,500-kilometer range with a 700-kilogram warhead, if it were to be used in a ballistic missile role.[42]

Iran has discussed a human space program, but it has turned out to be mostly bluster. In January 2013, Iran said it had successfully launched a monkey named Pishgam—Persian for pioneer—more than 70 miles up into space and then retrieved the animal alive, and the experiment was regarded by some as a prelude to human flight endeavors.[43] Elated over the success, President Ahmadinejad declared, "I'm ready to be the first Iranian to sacrifice myself for our country's scientists."[44] In May 2017, however, a semi-official Iranian news agency reported that the human space project had been canceled for cost reasons.[45]

Iran has a basic-level of indigenous satellite manufacturing and operations capability. Russia, Thailand, and China helped Iran develop and launch satellites into orbit during the latter half of the 2000s.[46] Using its own Safir space launch vehicle, Iran has launched four small satellites into orbit: Omid (2009), Rasad (2011), Navid (2012), and Fajr (2015).[47] These satellites were 50 kilograms or less and placed in such a low orbit that atmospheric drag caused them to reenter Earth's atmosphere within a few weeks. Additionally, it is believed that the low number of indigenous satellites in orbit may be as a result of sanctions or due to sensitivity of expected international reaction to launches because of their similar trajectories to ballistic missiles.[48] While Iran does have plans to launch larger satellites—both indigenous and in cooperation with other countries—those plans have seen recent delays.[49]

Based upon media reporting, some security analysts believe that Iran has demonstrated the ability to "spoof," or manipulate GPS signal information. In late 2011, it was reported that a U.S. RQ-170 Sentinel unmanned aerial vehicle

(UAV) landed erroneously in Iran.[50] The United States confirmed the event and subsequently asked for its return.[51] Some media reporting of the event suggests that Iranian specialists used a combination of techniques to misdirect the UAV, to include jamming the command and control signals and falsifying the "home base" GPS coordinates, so that the UAV landed in Iran and not at its home-base in Afghanistan.[52]

Other Iranian signal jamming includes that of satellite television signals. It was reported that in 2011, Iran increased its interference with the British Broadcast Channel, Voice of America, and other Western networks with Persian-language news channels.[53] Western broadcast media deemed the action as "intended to prevent Iranian audiences from seeing foreign broadcasts the Iranian government finds objectionable."[54] M.V. Smith has observed that "Iranian jamming of European satellite signals to prevent foreign news from entering Iran typifies the current state of space warfare."[55] Indeed, spoofing of GPS timing signals and satellite television communications are a proven method of impacting others.

Given past experience, Iran is likely to trudge along in its space program in an unspectacular manner. Doing so will realize some domestic benefit towards nationalist prestige. More importantly, anticipated actions will likely include investing in dual-use rocket technology and knowhow that can be employed for either launching satellites or ballistic missiles against would-be adversaries.

Non-military actions

As with great space powers, medium powers will have non-military means available to achieve political objectives. While there are various non-military means that medium powers may employ, four areas will be addressed here: diplomacy and alliances, buying power, establishing presence, and force in being.

Diplomacy and alliances

Diplomacy and the dialogue between states can help in advancing political objectives and achieving strategic ends. In describing the give and take that occurs during diplomacy, Thomas Schelling has commented,

> Diplomacy is bargaining; it seeks outcomes that, though not ideal for either party, are better for both than some of the alternatives. In diplomacy each party somewhat controls what the other wants, and can get more by compromise, exchange, or collaboration than by taking things in his own hands and ignoring the other's wishes.[56]

As a result, leaders of medium powers may employ diplomatic negotiation and compromise to protect and promote their space interests.

There is historical experience for this idea, as medium powers have participated in space-related international agreements or treaties through the United

Nations (UN) and other international bodies. Participation in international dialogue can assist a medium space power in shaping specific treaty language and regulations favorable to their state's interests. Even in cases when a medium space power cannot impose treaty language that is overtly self-serving, participation in international organizations can sometimes help ensure that resolutions or agreements that are directly harmful to a medium power's interests are not passed. An example of this advantage would be UN Security Council veto authority, such as that held by medium space powers France and the United Kingdom.

For medium space powers, international rules and regulations can influence access to and use of space. This thinking is highlighted in the 2016 "Space Strategy for Europe," which states, "Access to and use of space is shaped by international rules or standards and by a governance system aimed at guaranteeing the long-term, sustainable use of space for all nations."[57] The policy document states that "The EU, alongside its Member States and ESA, must act as a global stakeholder to promote and preserve the use of space for future generations. The EU cannot afford to fall behind in this domain."[58] As with many other policy and strategy documents, this one illustrates that European Union member states view that they can play an important role promoting common interests in space.

Diplomatic initiatives by medium space powers for a code of conduct have been proposed to help maintain the long-term sustainability, safety, stability, and security of space. A draft Code of Conduct for Outer Space Activities was published by the European Union in 2008, with a revised draft released in October 2014. A continuation of this effort led to a renaming to the International Code of Conduct, to address concerns raised by non-Europeans about the way the European Code of Conduct was coordinated and developed. Yet, even this repackaged initiative was ultimately unsuccessful and fell apart in July 2015 for a variety reasons and with numerous objections by countries. For example, India's space policy-makers expressed concern with the draft code's language, taking the view that to be workable, the final Code of Conduct required a legal framework, enforcement and verification mechanisms, along with penalty mechanisms for states violating the code.[59]

Even though a medium power will want to maintain its capability for independent action, cooperative relationships with other powers may be in its vital interests. For states, participation in alliances can help bolster one or more instruments of national power. For example, a group of like-minded medium space powers may have more diplomatic influence among the international community than any single state; therefore, forming a cooperative relationship among several states can improve diplomatic power and better promote common interests. Furthermore, a medium power may decide in favor of a cooperative relationship with a great power. The advantages of such an arrangement may include military protection in case of hostile action by a belligerent, through extended deterrence or mutual defense agreements. The downside of such a relationship includes any undue pressure on a medium power to be drawn into military actions into which the allied great power is involved and in which the medium power has little vested interest. Concerning the utility of alliances in addressing a medium power's threats to national security, J.R. Hill writes:

> Alliances, if structured, are to be based upon the help that could be expected from the ally or allies in the event of a threat to those interests. If ad hoc, alliance or coalitions are to be entered into on a judgment as to how a particular situation bears upon the vital interest of the nation-state.[60]

According to Hill, a medium power's strategy development should be interest-based. Threats—actual and potential—should be judged by the way they bear upon the medium power's vital interests.[61]

An example of a cooperative, collective self-defense agreement includes Article 5 of the Washington Treaty—the North Atlantic Treaty Organization's (NATO's) founding treaty. Article 5 describes that collective defense means that an attack against one ally is considered as an attack against all allies.[62] Of note, NATO invoked Article 5 for the first time in its history after the 9/11 terrorist attacks against the United States.[63] As borne by historical experience and the use of mutual defense treaties, being considered the non-aggressor and wrongfully attacked may allow others to come to a medium power's assistance, whether with diplomatic, economic, informational, or military means. In this way, a medium power can marshal additional alliance partners to defeat a common foe.

Buying power

Related to the economic instrument of national power is the concept of *buying power*. The idea is just merely seeking to convert one form of national power into another; in this case economic capability into a military one. For medium space powers looking to improve their space warfighting capability, procuring military capability through commercial means is a suitable action. Military history is rife with examples of states or regional powers buying the services of mercenaries to improve their military prowess. The second half of the twentieth century demonstrated a variation of this theme, where government leaders procured business and corporate services to augment military forces in providing logistical support and security forces. This was the case during the U.S. involvement in Iraq and Afghanistan in the 2000s, and Peter Singer referred to individuals performing such services as "corporate warriors."[64]

The same idea is applicable to a medium space power's space strategy. Medium space powers may enter into contracts, or service level agreements, with commercial space companies to augment or even be the sole provider of space-based services. Therefore, an economically well-off medium space power wanting to acquire military capability in space may be inclined to enter into a contract for this capability. One advantage of doing so is that a medium space power is not required to provide the research and development costs for any high-end systems upfront but only pay for specific services for a specified period. The services to be contracted may include those of dual-use nature, including on-orbit servicing and inspection services that can provide irreversible effects against satellites in orbit, as needed. Other potential contract services may include: remote sensing and associated data analytics; high-throughput satellite communications; heavy-lift launch vehicles; responsive launch capability; and extensive space situational awareness information.

A senior officer in the Israeli Air Force has similarly noted this idea of "buying" space access and capabilities. Whereas being a space power was formerly only considered relevant to those states with high-technology capability, "today commercial space technologies and capabilities are enabling smaller powers to have great access to space."[65] Utilizing commercial avenues to acquire space access and capabilities is thought to lower a state's risk when conducting military missions by providing operational advantage through additional capabilities.[66]

A potential disadvantage of a medium power contracting for effects of a military nature is that higher priority service level agreements may need to be honored, presuming the agreement was not of an exclusive nature. This may be the case when a medium and great space power are both competing for the same commercial services, and the more capable country will receive the higher priority and associated level of service, per the contract. Examples of contracted commercial services that are frequently in high demand during war include Earth imaging and wideband satellite communications services.

Presence

Establishing a noteworthy presence in space is necessary for a medium space power, if it has not already done so. Those with significant participation and presence in space activities achieve a proportionate level of influence in shaping international treaties and regulations. This advantage is similar to the benefits coming from prestige, as mentioned under Iran's space program. By increasing their participation in space-based and space-related activities, medium space powers are treated with a certain amount of respect and are given more consideration when contentious or competing issues arise with another space power. Only those having the highest levels of participation in space-based activities will achieve the greatest influence and positive results.

In noting the importance of presence in a medium power's maritime strategy and how presence can serve a broad spectrum of purposes, J.R. Hill writes:

> But usually, as service people well know, presence serves less well-defined objectives, demonstrating a variety of characteristics from fighting power at one end to intent of the most benign at the other. A telling characteristic of maritime forces is that they can cover the whole gamut at the same time. Moreover, medium-power maritime forces can do this without appearing to overbear or menace, as the forces of superpowers may too often do. Presence may also bring with it the opportunity to do really beneficent things:

disaster relief, search and rescue, projects for small and scattered communities.[67]

Commensurate with how other countries have used their navies, Hill observes that for medium powers, in general, the greater the visible fighting power the more influential the presence.[68]

For any military, numbers matter, and this fact plays into presence. Although the quality of military systems is always relevant during peace time deterrence operations and combat operations, the number of military space systems has intrinsic value in determining a state's relative standing to others. The actual number of assets or military platforms helps shape people's perceptions, because these numbers can be readily compared between states during quantitative comparison. Whether warranted or not, winning the quantitative comparison fight helps shape the perceptions of who is more powerful in space. This determination of presence, numbers, and capability will help in shaping the effectiveness of deterrence efforts—both deterrence through punishment and denial considerations.

Force in being

Related to the use of space presence to influence others is the *force in being* concept, which is derived from the "fleet in being" concept found in maritime strategy.[69] In maritime strategy, a less capable power should avoid a decisive military engagement against a stronger space power, and the less capable space force should be kept "in being" through active utilization and operation to achieve limited political ends until the situation improves in its favor. This idea will hold true for a medium space power, as well.

A medium space power can employ the concept of force in being to contest the command of another and advance national interests. By avoiding large-scale engagements with a superior space force, a lesser force can conduct minor, non-escalatory, frustrating, and harassing operations along celestial lines of communication (CLOCs) or against space-related activities, thus preventing a more capable power from gaining command of space that is either general or persistent. A medium power may do so at specific points along CLOCs to help establish local or temporary command of space, and efforts may be focused terrestrially or in space. Additionally, by using a force in being approach, and employing low-cost, expendable satellites in the process, a medium space power can mitigate the downside should tensions escalate and the space systems are destroyed. This idea is known in maritime strategy as a *disposal force*.[70]

Space strategist M.V. Smith writes on the implementation of a *force in being* concept when describing the utility of nonlethal methods in space. He observes that "... space weapons poised against uninhabited satellites constitute a nonlethal force 'in being.' Using such weapons in lieu of lethal means is in keeping with the spirit and intent of the law of armed conflict...."[71] Consequently, nonlethal methods—such as reversible, non-kinetic actions—can achieve the objectives of a force in being to achieve greater influence or political aims.

Chinese writings on space strategy discuss the role of a force in being, or influence through presence. A People's Liberation Army (PLA) document notes that, "Displays of space forces and weapons" may occur in either peacetime or at the onset of a crisis.[72] These displays serve to warn an opponent against escalating a crisis or pursuing a course of action that will lead to conflict.[73] Displays of space forces include using the media to highlight one's space capabilities, and these displays can be complemented further by political and diplomatic gestures and actions, such as inviting foreign military attachés to attend weapons tests and demonstrations.[74] According to PLA writings, if displays of force and weapons are insufficient to compel an adversary to alter its course, "military space exercises" may be conducted in peacetime or as a crisis escalates, to further influence an opponent's decision calculus.[75]

Offensive strategy and actions

The 2016 "Space Strategy for Europe" describes how space is of national importance to medium space powers in stating, "Space capacities are strategically important to civil, commercial, security and defence-related policy objectives. Europe needs to ensure its freedom of action and autonomy. It needs to have access to space and be able to use it safely."[76] The strategy document goes on to comment how space is becoming a "more contested and challenged environment," and "growing threats are also emerging in space...."[77] Implicit in the document's language, national interests need to be protected and any threats must be adequately addressed.

Indeed, the need to counter a determined aggressor and protect vital national interests are two reasons why offensive operations may at times be considered appropriate. According to Julian Corbett, the superior military power can usually employ offensive actions to obtain positive results while also attaining the "strength and energy" that comes from initiating attack.[78]

For a medium space power, this means offensive strategy can obtain positive results, in addition to boosting morale and imparting a psychological advantage to those forces initiating the attack. Moreover, the initiative gained through offensive operations may be beneficial because of the possibility of achieving operational surprise. In considering the utility of military forces in conducting various military and non-military focused missions, J.R. Hill reminds the strategist of the primary reason for having fighting forces in saying, "... navies are for fighting and [the strategist] must think how they should fight."[79] Likewise, while military assets in space can perform a variety of missions and functions, the strategist must think about their use during times of conflict.

Because medium powers will typically seek to act independently to protect national interests while also being constrained possibly by a shortage of fiscal and material resources, a medium space power's military services must be well versed in working as effectively and efficiently as possible. This effort includes the military services working well together or being "joint" in their combined use. By having land, sea, air, space, and cyber forces fully interoperable and well trained, a medium power's military capability will be better suited for carrying out its charter to act independently and for preserving vital interests.[80] Furthermore, space strategy must be viewed as a part of overall military strategy. Space strategy should not be understood as the sole means of achieving victory; consequently, space-based or space-enabled military actions are just another means available within the military instruments of power. For the above reasons, a medium power's space forces will need the support of the other military domains (land, sea, air, and cyber) working towards common military objectives, and the other military domains will likewise need the support of space forces to be as effective and efficient as possible.

Moreover, medium powers need a capable military force to enable effective deterrence. J.R. Hill notes fighting forces are for:

> convincing a potential opponent that military action against you will be unprofitable for him. So it is necessary to demonstrate the ability to fight in furtherance of vital national interests, and for that it is necessary to have forces that are ready, and effective, appropriately equipped and trained.[81]

In considering the role of space forces in deterring a country that is over-reliant on its space-based capabilities, M.V. Smith observes,

> It may be possible to deter an advanced spacefaring adversary who is heavily reliant on space systems but who has taken few or no precautions to defend them. In this case, possession of a credible set of offensive space weapons may cow the adversary into avoiding confrontations.[82]

The conclusion reached is medium powers can achieve some level of deterrence— given the military capability and specific conditions—even against those considered of a superior capability.

In cases when a medium power contests a great power, the medium power's preferred offensive strategy will usually require establishing command that is local or temporary in regions where its opponent is not. Temporary command will allow general or local command to be gained for specific periods to achieve either military or non-military objectives. Local command will allow temporary or persistent command to be established within a specific region. One can reasonably expect that a combination of actions and efforts at the strategic, operational, and tactical levels of war will be necessary ingredients when a medium power is establishing any level of command of space.

Limited war

Julian Corbett's idea of limited war highlights how using a smaller force against a larger one can achieve strategic advantage.[83] For Corbett, limited war is akin to the advantage enjoyed by the defense, which "sometimes enable an inferior force to gain its end against a superior one."[84] Historical experience shows how limited war has allowed naval powers to attain the initiative on both the strategic and tactical levels. Corbett describes how during the war with Napoleon, the use of small British amphibious forces to invade the continent or to divert the enemy forces to the coast "... was always out of all proportion to the intrinsic strength employed or the positive results it could give.... Its value lay in its power of containing [a] force greater that its own."[85] In countering the Clausewitzian view that the defeat of the enemy's ability to resist is always the primary object of offensive operations, Corbett argues:

> [T]he limited force of war has this element of strength over and above the unlimited form.... This point is of the highest importance, for it is direct negation of the current doctrine that there can be but one legitimate object, the overthrow of the enemy's means of resistance and that the primary objective must always be his armed forces.[86]

Using Corbett's idea of limited war, a medium space power may assume a limited offensive posture almost immediately to gain strategic advantage, without exposing itself to unacceptable risks.

Similar in sentiment but with a different emphasis, J.R. Hill writes on the utility of lower intensity operations for medium powers. Hill observes:

> The next level of conflict, as defined in medium-power strategy, is Low Intensity Operations. These can be defined as operations that never merit the title of war, are limited in aim, scope and area, are subject to the international law of self-defence, often include sporadic acts of violence by both sides, and have objectives that are predominantly political in nature.[87]

Hill notes that in more recent times, Low Intensity Operations involving democracies are likely to be multinational, rather than single-state, and "under the *nominal* aegis of a supranational organisation."[88] Low Intensity Operations may not require sizeable forces, and if the operation is multinational in participation, a single medium power's contribution may be small. On the other hand, if it is a single-state operation, then some careful force assessments will be necessary to address the situation and the potential opposition. Hill warns,

> Too much [force], against an indeterminate threat—perhaps from small bands of terrorists ashore, or harassing or quasi-piratical craft at sea and it will look like over-reaction; too little, and there is the possibility of an embarrassing casualty after a surprise attack.[89]

Consequently, the threat, political ends, and available forces all must be weighed when determining the desired action.

Thomas Schelling writes on the effect—whether intended or not—of limited war. Initiating or being part of limited conflict can provide a deterrent to continued aggression.[90] This is because conducting a limited war has a danger of expansion into an unlimited or major war.[91] Schelling writes, "To engage in limited war is to start rocking the boat, to set in motion a process that is not altogether in one's control."[92] Because limited war potentially raises the risk of escalation into a larger conflict, this potential consequence can also be a purpose for a limited war. Schelling writes:

> Deliberately raising the risk of all-out war is thus a tactic that may fit the context of limited war, particularly for the side most discontent with the progress of the war. Introduction of nuclear weapon undoubtedly needs to be evaluated in these terms.[93]

Because a main consequence of limited war is to potentially raise the risk of a larger war, medium space powers can use this fundamental point for deterrence's benefit.

Exploiting choke points

Medium space powers may seek military advantage by attacking an adversary's choke points. By attacking an enemy's CLOCs at choke points, a medium power can potentially have the greatest effect for the least amount of effort and expense. This thought is exemplified further by the pervasive use of space-based systems for military command and control that orchestrate various actions at the strategic, operational, and tactical levels of warfare. By denying or restricting the adversary's use of command and control communications at its orbital or terrestrial choke points, an adversary's ability to give timely orders can be severely limited, and thus affect his overall war fighting effectiveness. This is effectively impacting the adversary's orient, observe, decide and act (OODA) loop decision cycle. Examples of choke points include satellites or ground stations where a significant amount of data or communications are routed.

In addition to the seminal writings of Alfred Thayer Mahan, the idea of using choke points or other strategic positions for advantage can be found in the writings of the French naval officer Raoul Castex, who examined naval strategy for medium powers. In *Strategic Theories* [*Théories Stratégiques*], Castex applies the lessons of maritime strategy to those who were not sea powers.[94] Castex writes extensively on the use of geography for gaining strategic advantage over an adversary and states, "… war involving the attack and defense of communications is conditioned to the utmost degree by geography."[95] He viewed the influence of geography on maritime strategy as nothing other than the action of the land on the sea, and he highlighted that at times geography can offer defensive advantage.[96] Moreover, Castex discusses how geography's influence changes over time as available technology improves, and does so equally for everyone.[97] For him, a state's relative power depends largely on its physical configuration or geography.[98] Castex writes:

> The influence of geography on the general situation of communications has repercussions for the fleets because the number of forces that they have to

detach to attack the enemy communications and to defend their own will increase to the extent that geography places them at a disadvantage.[99]

Because a maritime theoretical framework is useful, at times, for considering space strategy, medium space powers can also think about the implications of "geography" when formulating space strategy. Indeed, practical experience has already done so. The "geography" of terrestrial launch locations influences how much energy is required to achieve orbit; the launch latitude helps define what orbital inclinations can be achieved easily. Geography helps define an antipodal choke point, which is a position through which each satellite must pass about a half revolution after its launch based on the antipode of its launch site.[100]

In additional to specific locations—albeit potentially moving—choke points may include certain orbital regimes. Because they are analogous to highly trafficked airways or sea lanes, some of the most desirable orbits have become regions more congested with satellites than other regions. The low Earth and geostationary orbital regions are two locations that have extensive activity, and therefore they could be considered "choke points" or regions of higher than normal activity. These two orbital regions are where about 90 percent of today's satellites operate.[101] If a medium space power was able to exploit the most congested of these orbital regions, while preserving its own use of them, a military advantage could be realized.

Dispersal and concentration

Due to its desire to act independently as well as having limited resources to meet all political objectives, a medium space power will likely need to employ the concept of dispersal and concentration to ensure that national interests are protected along vast CLOCs. Doing so will enable military effects to be concentrated where and when action is needed. This idea of dispersal and concentration reflects Castex's view of *manoeuvre*, referring to the capacity to move or act intelligently to create a favorable situation. Castex advocated for the most cost-effective use of naval forces, particularly when those naval forces could not dominate by sheer number or capability.[102]

Dispersal, as a general practice, also mitigates the likelihood of the adversary conducting a surprise attack against one's large concentration of forces, thereby reducing the chance that a foe can achieve military aims through a single decisive victory. Charles Callwell wrote on the advantages of dispersal and the threat of concentration when using amphibious forces against an adversary ashore. The enemy ashore must divide its forces, when a threating amphibious force has the option to "strike either to the left hand or the right."[103] Callwell goes on to say:

> The enemy is kept in a state of constant uncertainty. The hostile military forces have to be prepared for attack at many points. And the result of this is that the army of a nation which finds itself open to attack from the sea during the course of hostilities must of necessity be dispersed, and must to a

certain extent be scattered over the face of the territory which has to be defended.[104]

Callwell viewed dispersion as being married to concentration, because causing the enemy to scatter in the face of forces preparing to land from the sea made the adversary's forces ashore weak and vulnerable. Consequently, creating operational ambiguity can cause uncertainty in the enemy, thereby causing division in military forces.

If a medium space power disperses forces to the widest extent practicable, it will obtain those additional benefits coming from establishing presence. By moving and placing space systems and forces within a certain region, influence can be gained, and interests can be protected, even when actual force is not used. When the time comes for offensive actions, the concept of dispersal and concentration allows a medium space power to rapidly concentrate forces and effects against the enemy's decisive point to achieve the most successful results possible. Employing a strategy of dispersal and concentration preserves the flexibility of protecting expansive lines of communication while allowing an adversary's "central mass" to be engaged when needed.[105]

Although today's propulsive technology may prove limiting in physically moving between dispersal and concentration conditions, it is expected that technology will continue to mature and evolve to improve maneuverability in near-Earth orbit, cislunar space, and beyond. Furthermore, dispersal and concentration pertain to non-kinetic actions as well—including communications jamming, laser interference, and cyber actions against space-related infrastructure and networks. Such non-kinetic effects are readily understood through historical experience. Dispersal and concentration of a non-kinetic nature may include multiple systems—potentially in multiple domains—all acting in concert to impact negatively an adversary.

Commerce raiding

Drawn from centuries of maritime experience is the idea of commerce raiding, also referred to as *guerre de course*. The French navy employed this strategy when attacking British shipping and intercepting maritime commerce along trade routes. This maritime approach could be applied to space strategy because many states use CLOCs for business, trade and commerce, and medium space powers can affect negatively the economic interests of other space powers by impacting these lines of communication. An additional desired outcome of such action in space is to disturb the adversary's plans, while advancing the interests of the medium power. Because the relative standing and diplomatic effectiveness of a state results, in part, from its economic strength and the breadth of its commercial trade, upsetting space-related business, commerce and trade may affect the balance of power between competing states.[106] In any case, these kinds of actions are meant to achieve strategic effect at the expense of an adversary.

When the adversary is a weaker space power, the reason for employing this approach is to promote conditions that enable a quicker victory. When the adversary is another medium power, or even a great power, commerce raiding may be used to affect significantly the long-term sustainability of an opponent's attack. Attacking a peer power or great power's commerce and trade activities will foster conditions that may prolong a conflict allowing for the strategic element of time to be used in one's favor. Doing so may allow for conditions to turn in one's favor or at least delay eventual defeat.

While Corbett believed in the utility of commerce raiding in the maritime domain, he cautioned that the strategic effect of commerce raiding not be oversold. He warned that a singular focus on commerce raiding

> so often proved fatal and so often reborn as a new strategical discovery that a naval war may be conducted on economic principles and a great power be brought to its knees by preying on its commerce without first getting command of the sea.[107]

Similarly, Charles Callwell believed commerce raiding can provide benefit, but viewed it as depending upon the adversary. While the prosperity of the British Empire depended on the security of its maritime shipping, those nations whose wealth is not dependent on the sea cannot be similarly injured by the pressure of a dominating navy implementing a blockade of sea commerce. Callwell writes:

> The amount of damage which can be inflicted upon an antagonist by operations against his maritime trade obviously depends upon the volume of that trade. The value of the mercantile marine and the development of oversea commerce varies greatly in the case of different nations, and they are not necessarily proportionate to the importance or to the resources of a country as a whole.[108]

This lesson will hold for space strategy, as well. Affecting an adversary's space-reliant business, commerce, and trade can achieve strategic effect. Yet in doing so, it will depend on the adversary's reliance on space, along with a medium power establishing a significant level of power and influence—or command of space. Ultimately, commerce raiding is unlikely to decide by itself a conflict among impassioned belligerents.

Defensive strategy and actions

Defensive actions and preparations will be essential elements for a medium space power's space strategy. This condition is because, as Clausewitz advised, the defensive is the intrinsically stronger form of war.[109] Therefore, a defensive strategy is appropriate when a medium power is less capable militarily than an adversary—like against a great space power. A medium space power can use a defensive strategy to preserve its space capabilities and access to space, or the medium power may use such a strategy to prevent the enemy from acquiring something of value or achieving its political objectives.[110] If and when able, a medium power should consider when a defensive approach can be abandoned in order to pursue an offensive strategy.[111]

Guardians and resiliency

Medium space powers will have interests in space, and these interests may be considered vital. Consequently, medium space powers will need to develop and use systems that help ensure access to and use of space. Such systems may be called a variety of names, including *guardians* or *policing systems*.[112] Medium powers may decide to design and procure guardian systems that are inexpensive and numerous to disperse along the most vital CLOCs. The missions of these guardians may include those with defensive intent, to include patrolling CLOCs, escorting high-value assets, and self-defense actions as needed. If guardians can accomplish their intended mission objectives, medium powers may be better able to act independently in meeting their security needs.

Although guardians that help protect or defend critical space systems from hostile actions are important, it is also important to ensure continued access to and use of space after an attack occurs. As a result, some guardian systems should be specifically designed to include mission assurance or resiliency measures. By incorporating defensive preparations in such a manner, a medium power will be better able to perform vital space-enabled activities, even after a satellite is destroyed or essential CLOCs are attacked.

At present, CLOCs may utilize and transit multiple domains of operations. This observation implies that guardians and mission assurance measures must include land, sea, air, space, and cyberspace capabilities to be the most effective and efficient. The primary objective of space strategy is to ensure one's access to and use of space, and guardian systems—or whatever terminology is used—directly support this objective and are therefore of great importance in the conduct of space warfare.

Making space a "weak" barrier

A medium space power may use a defensive strategy to form a barrier against adversary action. This statement agrees with Clausewitz, who noted the usefulness of limited defensive war in saying: "The defender's purpose in the first category is to keep his territory inviolate, and to hold it for as long as possible. That will gain him time, and getting time is the only way he can achieve his aim."[113] Likewise, medium space powers may use defensive approaches to restrict or degrade another's access to and use of space to achieve limited aims. Doing so may help mitigate an adversary's space-enable effects within the other domains of conflict, delay defeat until the strategic situation changes, and contribute to conditions where an offensive strategy can be initiated.

Additionally, from an understanding of command of space and the concept of blocking, it is gleaned that space is readily accessible to those who exercise command; however, space becomes a "barrier" to those who cannot. Consequently, it is possible in some situations to make space a "weak" barrier that enables some limited objectives to be achieved. This approach can benefit medium powers because they are not likely to have resources to achieve command of space that is both general and persistent. A medium space power may have the ability to establish an adequate level of local command within a region, thereby locally establishing power and influence over other space powers. In doing so, a medium space power may be able to protect and defend its interests from attack, at least in the near-term.

A medium power may also duly consider the benefits coming from close blocking. Drawing upon maritime strategy, close blocking is obstructing or interfering with space communications in proximity to uplinks, downlinks, crosslinks, launching facilities, or any hubs of activity. Within Chinese space strategy writings, this idea is called "space blockading" activities, which include the blockade of terrestrial space facilities like launch sites; tracking, telemetry, and control sites; and mission control centers.[114] A close blocking strategy is more defensive in nature because it mostly seeks to prevent the enemy from action along CLOCs; and therefore, a medium space power may find it a suitable method to achieve limited aims. A close blocking strategy may be attempted when one adversary is weaker than its opponent, and the location chosen should be one with military advantage. As a result, a close blocking strategy is an appropriate option for a medium space power when the adversary is another medium power or even a great power.

Conclusion

A medium power is a state that prizes autonomy and is able to manipulate the instruments of national power in order to preserve itself. A medium space power will aim, consequently, to use space to enhance its ability to protect its interests.[115] Although it is understood that space will have a role in many spacefaring countries' national strategies, medium space powers will likely employ a space strategy different from those of emerging or great space powers. This contrast is due to a medium space power's desire to act independently, while likely being constrained by material and fiscal resources.

A number of medium space powers have established their prowess with regards to dual-use space technologies and capabilities—those that include both commercial and military utility. As these medium powers continue to develop military space capabilities, their respective space strategies will likely need to evolve to protect national interests and to address emerging security concerns. In the case of medium powers within the Indo-Pacific region, like India and Japan, this evolution may be especially true given that China is thought to be pursuing comprehensive military space capabilities, potentially including improved anti-satellite systems. Given the ever-changing global dynamics and growing relevance of space, the formulation of preferred space strategies for medium space powers is expected to remain relevant in shaping the international security environment in the years to come.

Notes

1 Deganit Paikowsky, "The Space Club—Space Policies and Politics" (paper presented at the 60th International Astronautical Congress, Daejeon, Republic of Korea, October 2009). Also referenced by Scott Pace, "A Space Launch without a Space Program," *38 North*, April 11, 2012, http://38north.org/2012/04/space041112

2 Ibid.

3 J.R. Hill, *Maritime Strategy for Medium Powers* (Annapolis, MD: Naval Institute Press, 1986), 20.

4 Ibid., 21. Italics are Hill's original emphasis.

5 J.R. Hill, "Medium Power Strategy Revisited," Working Paper No. 3 (Commonwealth of Australia: Royal Australian Navy, Sea Power Centre, 2000), 3, www.navy. gov.au/sites/default/files/documents/Working_Paper_3.pdf

6 Ibid.

7 Ibid.

8 Ibid., 26.

9 Ibid., 35. The context of Hill's comment is maritime strategy.

10 Geoffrey Till, *Seapower: A Guide for the Twenty-First Century*, 3rd ed. (Abingdon: Routledge, 2013), 151. Referenced in a quote attributed to Admiral Stansfield Turner.

11 Ibid., 148.

12 "Genesis," Indian Space Research Organization, accessed August 21, 2018, www. isro.gov.in/about-isro/genesis

13 "ISRO Home," Indian Space Research Organization, accessed August 21, 2018, www.isro.gov.in/

14 Narayan Prasad and Prateep Basu, "Renewing India's Space Vision: A Necessity or Luxury?" *The Space Review*, May 4, 2015, www.thespacereview.com/article/2742/1

15 Amrita Shah, "Flashback 1963: The Beginnings of India's Dazzling Space Programme; An Excerpt from Amrita Shah's 'Vikram Sarabhai—A Life', About the Father of India's Space Initiatives," *Scroll.In*, February 15, 2017, https://scroll.in/article/829466/flashback-1963-the-beginnings-of-indias-dazzling-space-programme; "ISRO's Timeline from 1960s to Today," Indian Space Research Organization, accessed August 21, 2018, www. isro.gov.in/about-isro/isros-timeline-1960s-to-today#1

16 "Satellites of India," Gunter's Space Page, accessed August 21, 2018, https://space. skyrocket.de/directories/sat_c_india.htm; "ISRO's Timeline from 1960s to Today."

17 Michael Sheehan, *The International Politics of Space* (Abingdon: Routledge, 2007), 152.

18 Rajeswari Pillai Rajagopalan, "Need For An Indian Military Space Policy," in *Space India 2.0: Commerce, Policy, Security and Governance Perspectives*, ed. Rajeswari Pillai Rajagopalan and Narayan Prasad (Observer Research Foundation, 2017), http://cf.orfonline.org/wp-content/uploads/2017/02/ORF_Space-India-2.0.pdf.

19 Joan Johnson-Freese, *Space Warfare in the 21st Century: Arming the Heavens* (Abingdon: Routledge, 2017), 47.

20 Dr. K. Kasturirangan, former head of the ISRO quoted in "Ex-ISRO chief calls China's A-SAT a cause for worry," *Press Trust of India*, September 14, 2009.

21 "ISRO Crosses 50 International Customer Satellite Launch Mark," Indian Space Research Organization, accessed August 21, 2018, www.isro.gov.in/isro-crosses-50-international-customer-satellite-launch-mark

22 Rajeswari Pillai Rajagopalan, "What's Next for India's Space Program?" *The Diplomat*, January 20, 2018, https://thediplomat.com/2018/01/whats-next-for-indias-space-program/

23 Ibid.

24 "Cartosat-2 Series Satellite," Indian Space Research Organization, last modified January 12, 2018, www.isro.gov.in/Spacecraft/cartosat-2-series-satellite-2

25 "Space Science & Exploration," Indian Space Research Organization, accessed August 21, 2018, www.isro.gov.in/spacecraft/space-science-exploration

26 "GSLV-F10/Chandrayaan-2 Mission," Indian Space Research Organization, accessed August 21, 2018, www.isro.gov.in/gslv-f10-chandrayaan-2-mission; Michael Roston, "Rocket Launches and Trips to the Moon We're Looking Forward To in 2018," *New York Times*, January 1, 2018, www.nytimes.com/2018/01/01/science/2018-spacex-moon.html

27 "Mars Orbiter Mission Spacecraft," Indian Space Research Organization, last modified November 5, 2013, www.isro.gov.in/Spacecraft/mars-orbiter-mission-spacecraft; Johnson-Freese, *Space Warfare in the 21st Century*, 37.

28 Johnson-Freese, *Space Warfare in the 21st Century*, 37.

29 Ibid., xvi.

30 Ibid.

31 Harsh Vasani, "India's Anti-Satellite Weapons: Does India Truly have the Ability to Target Enemy Satellites in War?" *The Diplomat*, June 14, 2016, http://thediplomat.com/2016/06/indias-anti-satellite-weapons/

32 Brian Harvey, Henk H.F. Smid, Théo Pirard, "Iran: Origins – the Road to Space," in *Emerging Space Powers: The New Space Programs of Asia, the Middle East, and South America* (Chichester, UK: Praxis Publishing, 2010), 256.

33 Ibid., 265.

34 Ibid.

35 Abolghasem Bayyenat, "Pride in the Future of Iran's Space Program," *Foreign Policy Journal*, July 7, 2011, www.foreignpolicyjournal.com/2011/07/07/pride-in-the-future-the-politics-of-irans-space-program/

36 Jassem Al Salami, "Iran Just Cancelled its Space Program," *War is Boring*, January 27, 2015, https://warisboring.com/iran-just-cancelled-its-space-program/

37 Brian Harvey, Henk H.F. Smid, Théo Pirard, "Iran: Development—Space Launch Systems and Satellites," in *Emerging Space Powers: The New Space Programs of Asia, the Middle East, and South America* (Chichester, UK: Praxis Publishing, 2010), 286.

38 Ibid.

39 Brian Weeden and Victoria Samson eds, "Global Counterspace Capabilities: An Open Source Assessment" (Secure World Foundation, April 2018), 4–3.

40 Bill Gertz, "Iran Conducts Space Launch," *Washington Free Beacon*, April 20, 2018, http://freebeacon.com/national-security/iran-conducts-space-launch/

41 "Iran Announces First Successful Simorgh Test Launch," *SpaceFlight101.com*, July 29, 2017, http://spaceflight101.com/iranannounces-first-successful-simorgh-test-launch/

42 Farzin Nadimi, "Iran's Space Program Emerges from Dormancy," Policywatch 2839 (The Washington Institute, August 1, 2017), www.washingtoninstitute.org/policy-analysis/view/irans-space-program-emerges-from-dormancy

43 William J. Broad, "Iran Reports Lofting Monkey into Space, Calling it Prelude to Human Flight," *New York Times*, January 28, 2013, www.nytimes.com/2013/01/29/world/middleeast/iran-says-it-sent-monkey-into-space.html

44 "Ahmadinejad Wants To Be First Iran Astronaut," *Aljazeera*, February 4, 2013, www.aljazeera.com/news/middleeast/2013/02/201324154448873605.html

45 Rick Gladstone, "Iran Drops Plan to Send Human into Space, Citing Cost," *New York Times*, May 31, 2017, www.nytimes.com/2017/05/31/world/middleeast/iran-space.html

46 Harvey, Smid, Pirard, "Iran: Development – Space Launch Systems and Satellites," 298–299.

47 Weeden and Samson, "Global Counterspace Capabilities," 4–3.

48 Ibid.

49 Ahmad Majidyar, "Iran Plans to Launch Several Satellites into Space, Including 1st Sensor-Operational Satellite" (Middle East Institute, May 30, 2017), www.mei.edu/

content/io/iran-plans-launch-several-satellites-space-including-1st-sensor-operational-satellite

50 Greg Jaffe and Thomas Erdbrink, "Iran Says It Downed U.S. Stealth Drone; Pentagon Acknowledges Aircraft Downing," *Washington Post*, December 4, 2011, www.washingtonpost.com/world/national-security/iran-says-it-downed-us-stealth-drone-pentagon-acknowledges-aircraft-downing/2011/12/04/gIQAyxa8TO_story.html

51 Rick Gladstone, "Iran is Asked to Return U.S. Drone," *New York Times*, December 12, 2011, www.nytimes.com/2011/12/13/world/middleeast/obama-says-us-has-asked-iran-to-return-drone.html

52 Scott Peterson and Payam Faramarzi, "Exclusive: Iran Hijacked U.S. Drone, Says Iranian Engineer," *Christian Science Monitor*, December 15, 2011, www.csmonitor.com/World/Middle-East/2011/1215/Exclusive-Iran-hijacked-US-drone-says-Iranian-engineer/

53 Paul Sonne and Farnaz Fassihi, "In Skies Over Iran, a Battle for Control of Satellite TV," *The Wall Street Journal*, December 27, 2011, www.wsj.com/articles/SB10001424052970203501304577088380199787036

54 Ibid.

55 M.V. Smith, "Spacepower and the Strategist," in *Strategy: Context and Adaption from Archidamus to Airpower*, eds. Richard J. Bailey Jr., James W. Forsyth Jr., and Mark O. Yeisley (Annapolis, MD: Naval Institute Press, 2016), 166.

56 Thomas C. Schelling, *Arms and Influence* (New Haven, CT: Yale University Press, 1966), 1.

57 European Commission, *Space Strategy for Europe*, Communication from the Commission to the European Parliament, the Council, the European Economic and Social Committee and the Committee of the Regions (October 26, 2016), 11.

58 Ibid., 13.

59 Rajeswari Rajagopalan, "Debate on Space Code of Conduct: An Indian Perspective," ORF Occasional Paper #26 (Observer Research Foundation, October 2011), 10, www.orfonline.org/research/debate-on-space-code-of-conduct-an-indian-perspective/

60 Hill, "Medium Power Strategy Revisited," 5.

61 Ibid.

62 "Collective Defence – Article 5," North Atlantic Treaty Organization, last modified June 12, 2018, www.nato.int/cps/ua/natohq/topics_110496.htm

63 "Collective Defence – Article 5."

64 P.W. Singer, *Corporate Warriors: The Rise of the Privatized Military Industry* (Ithaca, NY: Cornell University Press, 2003), 2–3.

65 Dani Haloutz, "Air and Space Strategy for Small Powers: Needs and Opportunities," in *Towards a Fusion of Air and Space: Surveying Developments and Assessing Choices for Small and Middle Powers*, eds. Dana J. Johnson and Ariel E. Levite (Santa Monica, CA: RAND Corporation, 2003), 148.

66 Ibid.

67 Hill, "Medium Power Strategy Revisited," 10–11.

68 Ibid., 19.

69 John J. Klein, *Space Warfare: Strategy, Principles and Policy* (Abingdon: Routledge, 2006), 122–123; Julian S. Corbett, *Some Principles of Maritime Strategy* (London: Longmans, Green and Co., 1911; reprint, Annapolis, MD: Naval Institute Press, 1988), 166. Corbett counters the "seek out and destroy" school of thought that was advocated by other strategists, such as Alfred Thayer Mahan.

70 Corbett, *Some Principles of Maritime Strategy*, 62.

71 Smith, "Spacepower and the Strategist," 168.

72 Dean Cheng, *Cyber Dragon: Inside China's Information Warfare and Cyber Operations* (Santa Barbara, CA: Praeger Security International, 2017), 166.

73 Ibid.

74 Ibid.

75 Ibid.
76 European Commission, *Space Strategy for Europe*, 8.
77 Ibid.
78 Corbett, *Some Principles of Maritime Strategy*, 34.
79 Hill, "Medium Power Strategy Revisited," 7.
80 Hill, *Maritime Strategy for Medium Powers*, 11, 20, 35.
81 Hill, "Medium Power Strategy Revisited," 8.
82 Smith, "Spacepower and the Strategist," 169.
83 Michael I. Handel, *Masters of War: Classical Strategic Thought*, 3rd ed. (London: Frank Cass, 2001), 293.
84 Corbett, *Some Principles of Maritime Strategy*, 74.
85 Ibid., 67.
86 Ibid., 74.
87 Hill, "Medium Power Strategy Revisited," 10.
88 Ibid.
89 Ibid., 11.
90 Schelling, *Arms and Influence*, 105.
91 Ibid.
92 Ibid., 105–106.
93 Ibid., 107.
94 Raoul Castex, *Strategic Theories*, trans. and ed. Eugenia C. Kiesling (Annapolis, MD: Naval Institute Press, 1994). As with many of his fellow naval officers, Castex wrote of the primacy of the battleship and described the sea as a great highway. Yet, unlike many naval strategists of his day, he wrote in depth on the role of geography on strategy.
95 Ibid., 280.
96 Ibid., 283.
97 Ibid., 280, 283.
98 Ibid., 283.
99 Ibid., 281.
100 James E. Oberg, *Space Power Theory* (Colorado Springs, CO: U.S. Space Command, 2000), 70.
101 Michael E. O'Hanlon, *Neither Star Wars nor Sanctuary: Constraining the Military Uses of Space* (Washington: Brookings Institute Press, 2004), 38.
102 Castex, *Strategic Theories*, xxxvi, xx.
103 Charles E. Callwell, *Military Operations and Maritime Preponderance: Their Relations and Interdependence*, with introduction by Colin S. Gray (London: William Blackwood and Sons, 1905; reprint, Annapolis, MD: Naval Institute Press, 1996), 267.
104 Ibid., 232.
105 Corbett, *Some Principles of Maritime Strategy*, 133. "Central mass" is the phrase used by Corbett, who was paraphrasing Clausewitz.
106 Ibid., 60.
107 As quoted in Till, *Seapower*, 214.
108 Callwell, *Military Operations*, 176–177.
109 Carl von Clausewitz, *On War*, trans. and eds. Michael Howard and Peter Paret (Princeton, NJ: Princeton University Press, 1989), 358.
110 Corbett, *Some Principles of Maritime Strategy*, 32.
111 Ibid.
112 Scott Pace, *Merchants and Guardians: Balancing U.S. Interests in Space Commerce* (Santa Monica, CA: RAND Corporation, 1999), www.rand.org/pubs/reprints/RP787.html; Klein, *Space Warfare*, 111–115.
113 Clausewitz, *On War*, 614.
114 Dean Cheng, "Evolving Chinese Thinking About Deterrence: What the United States Must Understand About China and Space," Backgrounder No. 3298 (The Heritage Foundation, March 29, 2018), 4–5, http://report.heritage.org/bg3298
115 This is a paraphrase of Hill, *Maritime Strategy for Medium Powers*, 48.

Bibliography

"Ahmadinejad Wants To Be First Iran Astronaut." *Aljazeera*. February 4, 2013. www.aljazeera.com/news/middleeast/2013/02/201324154448873605.html

Al Salami, Jassem. "Iran Just Cancelled its Space Program." *War is Boring*. January 27, 2015. https://warisboring.com/iran-just-cancelled-its-space-program/

Bayyenat, Abolghasem. "Pride in the Future of Iran's Space Program." *Foreign Policy Journal*. July 7, 2011. www.foreignpolicyjournal.com/2011/07/07/pride-in-the-future-the-politics-of-irans-space-program/

Broad, William J. "Iran Reports Lofting Monkey into Space, Calling it Prelude to Human Flight." *New York Times*. January 28, 2013. www.nytimes.com/2013/01/29/world/middleeast/iran-says-it-sent-monkey-into-space.html

Callwell, Charles E. *Military Operations and Maritime Preponderance: Their Relations and Interdependence*. With introduction by Colin S. Gray. London: William Blackwood and Sons, 1905; reprint, Annapolis, MD: Naval Institute Press, 1996.

"Cartosat-2 Series Satellite." Indian Space Research Organization. Last modified January 12, 2018, www.isro.gov.in/Spacecraft/cartosat-2-series-satellite-2

Castex, Raoul. *Strategic Theories*. Translated and edited by Eugenia C. Kiesling. Annapolis, MD: Naval Institute Press, 1994.

Cheng, Dean. "Evolving Chinese Thinking About Deterrence: What the United States Must Understand about China and Space." Backgrounder No. 3298. The Heritage Foundation, March 29, 2018. http://report.heritage.org/bg3298

Cheng, Dean. *Cyber Dragon: Inside China's Information Warfare and Cyber Operations*. Santa Barbara, CA: Praeger Security International, 2017.

Clausewitz, Carl von. *On War*. Translated and edited by Michael Howard and Peter Paret. Princeton, NJ: Princeton University Press, 1989.

"Collective Defence – Article 5." North Atlantic Treaty Organization. Last modified June 12, 2018. www.nato.int/cps/ua/natohq/topics_110496.htm

Corbett, Julian S. *Some Principles of Maritime Strategy*. London: Longmans, Green and Co., 1911; reprint, Annapolis, MD: Naval Institute Press, 1988.

European Commission. *Space Strategy for Europe*. Communication from the Commission to the European Parliament, the Council, the European Economic and Social Committee and the Committee of the Regions. October 26, 2016.

"Ex-ISRO Chief Calls China's A-SAT a Cause For Worry." *Press Trust of India*. September 14, 2009.

"Genesis." Indian Space Research Organization. Accessed August 21, 2018. www.isro.gov.in/about-isro/genesis

Gertz, Bill. "Iran Conducts Space Launch." *Washington Free Beacon*. April 20, 2018. http://freebeacon.com/national-security/iran-conducts-space-launch/

Gladstone, Rick. "Iran Drops Plan to Send Human into Space, Citing Cost." *New York Times*. May 31, 2017. www.nytimes.com/2017/05/31/world/middleeast/iran-space.html

Gladstone, Rick. "Iran is Asked to Return U.S. Drone." *New York Times*. December 12, 2011. www.nytimes.com/2011/12/13/world/middleeast/obama-says-us-has-asked-iran-to-return-drone.html

"GSLV-F10/Chandrayaan-2 Mission." Indian Space Research Organization. Accessed August 21, 2018. www.isro.gov.in/gslv-f10-chandrayaan-2-mission

Haloutz, Dani. "Air and Space Strategy for Small Powers: Needs and Opportunities." In *Towards a Fusion of Air and Space: Surveying Developments and Assessing Choices for Small and Middle Powers*, edited by Dana J. Johnson and Ariel E. Levite, 147–157. Santa Monica, CA: RAND Corporation, 2003.

Handel, Michael I. *Masters of War: Classical Strategic Thought*. 3rd edition. London: Frank Cass, 2001.

Harvey, Brian, Henk H.F. Smid, and Théo Pirard. "Iran: Development – Space Launch Systems and Satellites." In *Emerging Space Powers: The New Space Programs of Asia, the Middle East, and South America*, 285–350. Chichester, UK: Praxis Publishing, 2010.

Harvey, Brian, "Iran: Origins – the Road to Space." In *Emerging Space Powers: The New Space Programs of Asia, the Middle East, and South America*, 255–284. Chichester, UK: Praxis Publishing, 2010.

Hill, J.R. "Medium Power Strategy Revisited." Working Paper No. 3. Commonwealth of Australia: Royal Australian Navy, Sea Power Centre, 2000. www.navy.gov.au/sites/default/files/documents/Working_Paper_3.pdf

Hill, J.R. *Maritime Strategy for Medium Powers*. Annapolis, MD: Naval Institute Press, 1986.

"Iran Announces First Successful Simorgh Test Launch." *SpaceFlight101.com*. July 29, 2017. http://spaceflight101.com/iranannounces-first-successful-simorgh-test-launch/

"ISRO Crosses 50 International Customer Satellite Launch Mark." Indian Space Research Organization. Accessed August 21, 2018. www.isro.gov.in/isro-crosses-50-international-customer-satellite-launch-mark

"ISRO Home." Indian Space Research Organization. Accessed August 21, 2018. www.isro.gov.in/

"ISRO's Timeline from 1960s to Today." Indian Space Research Organization. Accessed August 21, 2018. www.isro.gov.in/about-isro/isros-timeline-1960s-to-today#1

Jaffe, Greg and Thomas Erdbrink. "Iran Says It Downed U.S. Stealth Drone; Pentagon Acknowledges Aircraft Downing." *Washington Post*. December 4, 2011. www.washingtonpost.com/world/national-security/iran-says-it-downed-us-stealth-drone-pentagon-acknowledges-aircraft-downing/2011/12/04/gIQAyxa8TO_story.html.

Johnson-Freese, Joan. *Space Warfare in the 21st Century: Arming the Heavens*. Abingdon: Routledge, 2017.

Klein, John J. *Space Warfare: Strategy, Principles and Policy*. Abingdon: Routledge, 2006.

Majidyar, Ahmad. "Iran Plans to Launch Several Satellites into Space, Including 1st Sensor-Operational Satellite." Middle East Institute, May 30, 2017. www.mei.edu/content/io/iran-plans-launch-several-satellites-space-including-1st-sensor-operational-satellite

"Mars Orbiter Mission Spacecraft." Indian Space Research Organization. Last modified November 5, 2013. www.isro.gov.in/Spacecraft/mars-orbiter-mission-spacecraft

Nadimi, Farzin. "Iran's Space Program Emerges from Dormancy." Policywatch 2839. The Washington Institute, August 1, 2017. www.washingtoninstitute.org/policy-analysis/view/irans-space-program-emerges-from-dormancy

Oberg, James E. *Space Power Theory*. Colorado Springs, CO: U.S. Space Command, 2000.

O'Hanlon, Michael E. *Neither Star Wars nor Sanctuary: Constraining the Military Uses of Space*. Washington: Brookings Institute Press, 2004.

Pace, Scott. "A Space Launch Without a Space Program." *38 North*. April 11, 2012. http://38north.org/2012/04/space041112

Pace, Scott. *Merchants and Guardians: Balancing U.S. Interests in Space Commerce*. Santa Monica, CA: RAND Corporation, 1999. www.rand.org/pubs/reprints/RP787.html

Paikowsky, Deganit. "The Space Club—Space Policies and Politics," Paper presented at the 60th International Astronautical Congress. Daejeon, Republic of Korea, October 2009).

Peterson, Scott and Payam Faramarzi. "Exclusive: Iran Hijacked U.S. Drone, Says Iranian Engineer." *Christian Science Monitor*. December 15, 2011. www.csmonitor.com/

World/Middle-East/2011/1215/Exclusive-Iran-hijacked-US-drone-says-Iranian-engineer/

Prasad, Narayan and Prateep Basu. "Renewing India's Space Vision: A Necessity or Luxury?" *The Space Review*. May 4, 2015. www.thespacereview.com/article/2742/1

Rajagoplan, Rajeswari Pillai. "What's Next for India's Space Program?" *The Diplomat*. January 20, 2018. https://thediplomat.com/2018/01/whats-next-for-indias-space-program/

Rajagoplan, Rajeswari Pillai. "Need For An Indian Military Space Policy." In *Space India 2.0: Commerce, Policy, Security and Governance Perspectives*, edited by Rajeswari Pillai Rajagopalan and Narayan Prasad, 199–214. Observer Research Foundation, 2017. http://cf.orfonline.org/wp-content/uploads/2017/02/ORF_Space-India-2.0.pdf

Rajagoplan, Rajeswari Pillai. "Debate on Space Code of Conduct: An Indian Perspective." ORF Occasional Paper #26. Observer Research Foundation, October 2011. www.orfonline.org/research/debate-on-space-code-of-conduct-an-indian-perspective/

Roston, Michael. "Rocket Launches and Trips to the Moon We're Looking Forward To in 2018." *New York Times*. January 1, 2018. www.nytimes.com/2018/01/01/science/2018-spacex-moon.html

"Satellites of India." Gunter's Space Page. Accessed August 21, 2018. https://space.skyrocket.de/directories/sat_c_india.htm

Schelling, Thomas C. *Arms and Influence*. New Haven, CT: Yale University Press, 1966.

Shah, Amrita. "Flashback 1963: The Beginnings of India's Dazzling Space Programme; An Excerpt from Amrita Shah's 'Vikram Sarabhai – A Life', About the Father of India's Space Initiatives." *Scroll.In*. February 15, 2017. https://scroll.in/article/829466/flashback-1963-the-beginnings-of-indias-dazzling-space-programme

Sheehan, Michael. *The International Politics of Space*. Abingdon: Routledge, 2007.

Singer, P.W. *Corporate Warriors: The Rise of the Privatized Military Industry*. Ithaca, NY: Cornell University Press, 2003.

Smith, M.V. "Spacepower and the Strategist." In *Strategy: Context and Adaption from Archidamus to Airpower*, edited by Richard J. Bailey Jr., James W. Forsyth Jr., and Mark O. Yeisley, 157–185. Annapolis, MD: Naval Institute Press, 2016).

Sonne, Paul and Farnaz Fassihi. "In Skies Over Iran, a Battle for Control of Satellite TV." *The Wall Street Journal*. December 27, 2011. www.wsj.com/articles/SB10001424052970203501304577088380199787036

"Space Science & Exploration." Indian Space Research Organization. Accessed August 21, 2018. www.isro.gov.in/spacecraft/space-science-exploration

Till, Geoffrey. *Seapower: A Guide for the Twenty-First Century*. 3rd edition. Abingdon: Routledge, 2013.

Vasani, Harsh. "India's Anti-Satellite Weapons: Does India truly have the Ability to Target Enemy Satellites in War?" *The Diplomat*. June 14, 2016. http://thediplomat.com/2016/06/indias-anti-satellite-weapons/

Weeden, Brian and Victoria Samson, editors. "Global Counterspace Capabilities: An Open Source Assessment." Secure World Foundation, April 2018.

7 Space strategy for emerging powers

This chapter discusses strategy considerations for emerging space powers. The space programs of Canada and Saudi Arabia are examined, along with historical experience of the space activities of non-state actors. Then, this chapter addresses the topics of diplomatic initiatives, dispersal and concentration, asymmetric operations, and protracted warfare. Also, potential terrorist actions related to space activities are addressed.

As with medium powers, emerging powers are thought to be less capable when compared to great space powers. Even though, having limited means, the least capable may still decide to exploit space or space-related activities for strategic advantage to achieve political ends. When formulating space strategy and considering military operations, Mao Tse-tung's advice is eternal: "All the guiding principles of military operations grow out of the one basic principle: to strive to the utmost to preserve one's own strength and destroy that of the enemy."[1] Hence, the preservation of limited forces and assets should always be at the forefront of considerations for emerging space powers.

Using the prior category definitions for those part of the "space club," emerging space powers include the numerous states that indigenously develop, maintain, and control satellites, but who are unable to launch them through indigenous means.[2] Yet to consider space strategy more broadly, this definition should be expanded somewhat. Although space strategy is usually considered within the context of state actors, non-state actors—including non-governmental organizations, corporations, insurgencies, and terrorist groups—may also seek to pursue space-related interests for either strategic or political ends. Consequently, non-state actors should also be included when describing the potential actions for those less capable space powers.

Dave Baiocchi and William Welser have described situations where both state and non-state actors will be involved as "the new space race."[3] The authors write that many of the smaller and less expensive space missions of tomorrow will be funded by cross-national teams and private interests, and it will become harder to both assess the intent of a particular mission and assign any associated liability to the responsible party under the Outer Space Treaty in the event of a mishap.[4] New space players may feel free to operate independent of national policies. As a result, non-state actors may pursue strategic or political ends that undermine a government's national-level objectives. Similar to this idea of the new space race, the 2017 U.S. National Security Strategy refers to the increased role of the private sector and other "motivated actors" as the *democratization of space.*[5] The *democratization of space* means many governments and non-state actors can launch satellites into space at relatively low cost.[6] According to the strategy document, the democratization of space and the ability to exploit the fusion of data from imagery, communications, and geolocations services can impact adversely U.S. military operations and America's ability to prevail in conflict.[7]

In general, the strategy for emerging powers appears to be an underdeveloped area of strategy when compared to those considered super or great powers. Despite this situation, it is still possible to glean useful ideas from the writings on the strategy for insurgencies. The strategies for insurgencies, or *popular uprisings* as referred to in Clausewitz's chapter titled "The People in Arms," are relevant because insurgencies can be considered a sub-category of emerging powers.[8] *Insurgents* are often thought of as a group seeking some political goal, which commonly includes autonomous self-governance, using a protracted strategy. The term *guerrilla warfare* is also commonly used when thinking about operational actions by insurgents.[9]

In noting the limited amount of writings on the subject of insurgencies, Clausewitz writes, "This discussion has been less an objective analysis than a groping for the truth."[10] Even though not being well understood or widely written about, he notes that this type of warfare will include "the elemental violence of war," as with more traditional styles of warfare.[11] While Clausewitz's observation on the scarcity of fulsome insurgent strategies also holds true for the strategy of emerging powers in general, war and warfare have an enduring nature. Consequently, it is possible to glean general strategy considerations for the less capable from the writings of Sun Tzu, Clausewitz, Mao Tse-tung, B.H. Liddle Hart, and J.C. Wylie. These five strategists cover many enduring ideas that remain relevant when considering the strategies—including space strategies—of emerging powers.

Before detailing the framework for emerging space powers, it is helpful to provide a few examples. In the end, any theoretical strategic framework should be useful in its application, and therefore, grounding these ideas in a practical understanding is beneficial. In discussing the strategy for emerging space powers, the following sections will first describe the space activities of Canada and Saudi Arabia. These two were chosen because the former has a long history of space activity and the latter is a more recently established emerging space power. Second, some broad strategy considerations and potential objectives for emerging powers will be discussed to provide a framework for thinking about a space strategy. Third, a range of potential non-military actions will be described. Fourth, a range of military approaches will be considered, with the understanding that an emerging space power's options will likely differ out of necessity from those considered more capable. Last, potential actions by terrorist organizations will be addressed, because terrorists are a sub-group of emerging powers and will likely use methods different than others.

Canada

Despite being considered an emerging space nation according to the definition used, Canada has a long and notable history of space activity. The launch of Canada's first satellite, Alouette-I, in 1962 made Canada the fourth nation to operate a satellite in orbit, after the Soviet Union, United States, and United Kingdom respectively.[12] Alouette-I was a scientific satellite and was launched on top of an American Thor-Agena vehicle, which highlights the beginning of a close relationship between Canada and the United States in space exploration.[13] The development of Alouette-I came as a result of an American invitation, through the newly formed National Aeronautics and Space Administration (NASA) in 1958, for international collaboration. Following the success of Alouette-I, Canada and the United States signed an agreement to launch additional satellites under a new program called International Satellites for Ionospheric Studies.[14] Today, Canada's focus on space is to support science and exploration, support the private sector, continue to invest in key capabilities, and to serve as an inspiration for Canadians.

Canada does not have an indigenous space launch capability, although it has developed Black Brant sounding rockets, which have become one of the most popular rockets of their kind.[15] However, relationships with both NASA and the European Space Agency (ESA) have allowed Canada to use the launch systems of other countries and focus their space efforts elsewhere.[16] Canada's advanced Canadarm technology is such an example. The Canadian technology is considered by many to be state-of-the-art, and it was used on the U.S. space shuttle as the Shuttle Remote Manipulator System and was a critical capability in most of the Shuttle missions.[17] The Canadarm-2, a larger and more capable system, has been extensively involved in assembly of the International Space Station since 2001.[18]

Established in 1990, the Canadian Space Agency (CSA) has also played an important role in many other signature NASA missions. NASA's Curiosity Rover, which currently is exploring Mars to determine whether the Red Planet ever had the conditions to support life, carries a Canadian-made Alpha Particle X-ray Spectrometer. The instrument allows Curiosity to read the chemical composition of the Martian soil and rocks.[19] CSA has also provided a series of components for the forthcoming James Webb Space Telescope, which is an international collaboration between NASA, ESA, and CSA.[20]

A total of nine Canadians have been to space, with David Saint Jacques scheduled to be the tenth in December 2018 for International Space Station (ISS) Expedition 58/59.[21] In 1984, the first Canadian astronaut Marc Garneau reached space on board the *Challenger* STS-41-G mission as a payload specialist. After further training, he served as mission specialist aboard *Endeavor* STS-77 and STS-97 missions.[22] Canada's most famous astronaut is Chris Hadfield. He served as a Mission Specialist aboard NASA's STS-74 and STS-100 missions and later became the first Canadian to command the ISS during Expedition 35 in 2013. His social media popularity has helped increase the popularization of space exploration to the general public, specifically for a new generation of Canadians interested in space.[23]

Despite Canada's long history in space, it did not launch a dedicated military satellite until February 2013, over 50 years after Alouette-I. The Canadian Armed Forces' satellite, named Sapphire, was launched in Sriharikota, India by the Indian Space Research Organization.[24] The primary contractor, MacDonald, Dettwiler and Associates Ltd., developed and built the satellite, and its mission is to collect observations of objects orbiting between 4,000 and 6,000 kilometers above the Earth using an electro-optical sensor. Sapphire is designed to be a key element of the Canadian space surveillance system, along with contributing to the U.S. Space Surveillance Network.[25]

Saudi Arabia

Countries within the Middle East have emerged seeking access to and use of space, to include Saudi Arabia. While Saudi Arabia is a prominent and influential actor within the region, its level of space capability places it in the category of an emerging space power. To date Saudi Arabia's use of space has been relatively modest but, geopolitical developments in its neighborhood have prompted the kingdom to reconsider the benefits of maintaining satellites in orbit.

Saudi Arabia's spacefaring history began in the form of a cooperative venture. The Arab Satellite Communications Organization (Arabsat), an intergovernmental organization and headquartered in Riyadh, was founded in 1976 by the member-states of the Arab League with the purpose of reaching millions of Arabic speaking viewers and providing telecommunications services.[26] Saudi Arabia became the main financier of Arabsat and continues to contribute most of the organization's capital—more than double the next highest contributor.[27]

Saudi Arabia has limited experience in operating satellites. Arabsat 1A, contracted from Aerospatiale and Messerschmitt-Bölkow-Blohm and launched on an Ariane rocket in 1985, suffered a solar panel malfunction immediately after launch.[28] After this initial setback, Arabsat 1B was launched later that same year. The satellite launched onboard the U.S. Space Shuttle *Discovery*. Arabsat 1B was deployed successfully by a Shuttle crew that included Sultan bin Salman Al Saud, second son of King Salman, as a payload specialist.[29] Prince Sultan became the first Saudi Arabian, Arab, Muslim, and royal to be in space. He became a national icon and was a source of inspiration for young Arabs across the region.[30] Since 1985, Arabsat has operated or leased the services of over six series of Arabsat satellites, encompassing over 15 satellites. Arabsat 6A, set to launch in 2019, is expected to be the first commercial satellite to launch onboard SpaceX's Falcon Heavy launch vehicle.[31]

In addition to its engagement and cooperation with Arabsat, Saudi Arabia has pursued indigenous satellite manufacturing capabilities. Most of Saudi Arabia's space efforts are housed within the King Abdulaziz City for Science and Technology (KACST). The Saudi satellite program started in 1998 with the establishment of the Space Research Institute at KACST. With a goal to produce the human capital and infrastructure needed for greater space endeavors, KACST produced Saudi Arabia's first satellites, the two microsatellites Saudisat 1A and 1B, which launched onboard the Russian Dnepr rocket.[32] KACST continued developing Saudisat 2 and 3 for Earth observation missions and during the mid-2000s, KACST produced the seven SaudiComsat micro-communications satellites.[33] Seven years after the launch of the last series SaudiComsats, Saudisat 4 was launched in 2014 onboard a Russian launch vehicle.[34] Saudisat 5B launched in 2018 onboard a Long March 2D from Jiuquan Satellite Launch Center. In addition to developing satellites, KACST is building state-of-the-art satellite testing facilities and plans to develop an advanced ground station for telemetry, tracking and command of its constellation of satellites.[35]

Saudi Arabia's role as an emerging space power has prompted it to partner and cooperate with more established spacefaring nations and private companies. KACST is pursing partnerships with France's Centre National d'études Spatiales to facilitate data and personnel exchange on satellite remote sensing, space science, small satellite development, and space regulations.[36] Saudi Arabia's engagement with China has led to a Saudi payload camera being incorporated into China's Longjiang-2 lunar microsatellite, which is a preparatory mission for China's Chang'e-4 lunar mission.[37]

The Saudi government is pursuing cooperation with commercial companies. For example, KACST, Taqnia Space—a company owned by Saudi Arabia's Public Investment fund—and DigitalGlobe formed a joint venture in which KACST will build six Earth observation satellites with sub 1-meter resolution to complement DigitalGlobe's constellation of satellites.[38] As part of the agreement, KACST will own 50 percent of the satellites' capacity over Saudi Arabia and the surrounding area, while DigitalGlobe will retain the rest. In another example, Taqnia Space has signed an agreement with Lockheed Martin to build a satellite assembly plant on Saudi soil.[39] Saudi Arabian investments in the commercial launch sector includes $1 billion investment in the Virgin Group's sub-orbital and orbital space launch vehicles.[40]

Many of the recent developments in the Saudi space sector and its growing embrace of space power appear to stem from Prince Mohammed bin Salman, the son of King Salman and Crown Prince of Saudi Arabia. Prince Mohammed has developed an ambitious economic and social reform plan called Vision 2030.[41] As the kingdom attempts to diversify its oil dependent economy and maintain its prominence in the region, it is thought that space is a suitable sector to bolster the Saudi economy and help develop a skilled professional workforce that includes scientists, engineers, and technicians, while also encouraging younger Saudi Arabians to pursue careers in science, technology, engineering, and mathematics fields.[42]

Non-state actors

When describing interests in space, it is not only countries that have equities, but also non-state actors. In noting this, the 2018 U.S. National Defense Strategy provides context on non-state actors.[43]

States are the principal actors on the global stage, but *non-state actors* also threaten the security environment with increasingly sophisticated capabilities. Terrorists, trans-national criminal organizations, cyber hackers and other malicious non-state actors have transformed global affairs with increased capabilities of mass disruption.[44]

When considering a sound space strategy, non-state actors should be included, as their efforts may touch on political or strategic ends, along with any associated means to achieve them.

For the sake of brevity, not all non-state actors can be considered here. Regardless, it is important to realize that companies and intergovernmental consortiums may have interests that need to be advanced. Commercial space companies and related topics will be covered in Chapter 8, but in general, the strategic objectives of companies may include increased market share, revenue, and innovation. In some cases, as with the aspirational goals of Jeff Bezos and Elon Musk, the strategic goals may be to establish a space tourism industry or the colonization of Mars.[45]

Similarly, an international consortium may have high-level objectives. Two examples of international consortiums are found in the histories of Intelsat and Inmarsat. The company Intelsat was originally formed as the International Telecommunications Satellite Organization, and from 1964 to 2001, it operated as an intergovernmental consortium owning and managing a constellation of communications satellites providing international broadcast services.[46] The company Inmarsat was established in 1979 as a non-profit intergovernmental organization but has since become privatized, providing telephone and data services to users worldwide via portable or mobile terminals using a constellation of geostationary telecommunications satellites.[47] The objectives of international consortiums may include establishing standards among the industry, norms of behavior for use of spectrum or on-orbit operations, promoting public safety through the use of satellite-derived data and information, and improving the scientific understanding and technological capabilities.

Cumulative strategy and the indirect approach

Two specific concepts are illuminating when discerning potential strategies for emerging powers: a cumulative strategy, as described by J.C. Wylie, and the indirect approach, as described by B.H. Liddell Hart. Even though the concepts of cumulative strategy and indirect approach are also applicable to great and medium powers, they are considered especially relevant to emerging powers. Also, while writings on the regular style of warfare do not exclude these concepts, seldom are these ideas emphasized as much as they should be. The indirect approach and cumulative strategy will likely be centerpieces for an emerging power's space strategy, as demonstrated by historical experience and the formulation of insurgent warfare strategy.

In writing about the various operational styles of non-military and military activities, Wylie saw psychological warfare, economic warfare, naval blockades, and guerrilla warfare as operations and activities where a cumulative strategy is typically employed.[48] According to Wylie, a cumulative strategy uses the "minute accumulation of little items piling on top of the other" until the mass of accumulated actions becomes critical.[49] For example in a naval blockade, while a few opponent's ships might make it through, the overall effectiveness of the blockade may not be compromised; in guerrilla warfare, an individual platoon of insurgents might be discovered and defeated, but the overall insurgency's effort remains intact as long as other guerrilla forces remain, believe in the cause, and continue the fight.

According to Wylie, the downside of a cumulative strategy is that it is rarely enough to defeat the enemy and achieve the requisite level of control to conclude decisively a conflict. He gives the example of this inconclusive ability as the French use of *guerre de course* at sea—or commerce raiding.[50] Impacting negatively an adversary's commerce and trade on the seas may contribute to eventual success, but it is difficult for it to be the sole determinant in doing so. However, when both sequential (a series of discrete actions that depend upon the one that precede it) and cumulative strategies are combined into an integrated effort, maximum pressure on the enemy is achieved to establish control over him.[51] Wylie viewed historical experience as demonstrating many occasions when the strength of the cumulative strategy has indeed meant the difference between success and failure of a sequential strategy.[52]

Wylie believed in the soundness of Mao Tse-tung's strategy of guerrilla warfare, or, as he also termed, the "war of national liberation."[53] Wylie saw guerrilla actions as having a long, established history and noted that there was, in fact, no novelty in this type of warfare.[54] More importantly, he viewed Mao's theory of conflict as exemplifying many of his ideas on cumulative strategy. Wylie writes, "It is important to note here that while the normal strategy of the continental or Clausewitz theory is a sequential strategy, a main weight of the Mao theory is based on a cumulative, not sequential, concept."[55] In commenting on the importance of Mao and others' theory with respect to wars of national liberation, Wylie held that the actual practice has been successful, and therefore, the books and theory on the subject should be of immeasurable importance to every strategist—whether military or civilian—in Western society today. He writes, "These books are not only theory, they portray a hard reality of contemporary warfare."[56] Heeding Wylie's advice, Mao's theory of warfare is postulated on reality; consequently, guerrilla theory must be included within the general theories of warfare.[57]

It is noteworthy that Wylie wrote on the applicability of B.H. Liddell Hart's writing in guerrilla warfare. Wylie comments that Liddell Hart's idea of the indirect approach is "... much more receptive to the concepts of Mao than those of Clausewitz."[58] This belief is because Liddell Hart's *indirect approach* incorporates the thought that strategy should adjust as the situation develops in war. In writing to counter what he saw was an incorrect interpretation of Clausewitz's theory, Liddell Hart explained how strategy should not seek solely to overcome the adversary's resistance, but rather should exploit the elements of movement and surprise to achieve operational advantage by throwing the enemy off balance before a potential strike.[59]

Mao's writings do, in fact, incorporate the ideas of cumulative strategy and the indirect approach—even though he does not reference them necessarily in those terms. Regarding the idea of the cumulative strategy and need for successes to be accrued together, Mao writes:

> If each month we could win one sizable victory like that at Pinghsingkuan or Taierhchuang, not to speak of more, it would greatly demoralize the enemy, stimulate the morale of our own forces and evoke international support. Thus our strategically protracted war is translated in the field into battles of quick decision. The enemy's war of strategic quick decision is bound to change into protracted war after he is defeated in many campaigns and battles.[60]

In contrast, with respect to the indirect approach and the need to be adaptable to the situation at hand, Mao writes:

> Flexibility in dispersal, concentration and shifts of position is a concrete expression of the initiative in guerrilla warfare, whereas rigidity and inertia inevitably lead to passivity and cause unnecessary losses. But a commander proves himself wise not just by recognition of the importance of employing his forces flexibly but by skill in dispersing, concentrating or shifting them in good time according to the specific circumstances.[61]

Because the conditions of war are constantly changing, the execution of the war plan must evolve as well. This evolution is exemplified in Mao's statement, "War is a contest of strength, but the original pattern of strength changes in the course of war."[62] Part of this need to be flexible includes avoiding the enemy's strength and striking its weakness, which agrees with the indirect approach. By creating situations that offer clear advantage through a thoughtful and sound strategy, a military can potentially realize an operational advantage.

Potential strategic objectives

If using a rational actor model of international affairs, an emerging space power may weigh the potential risks and rewards of any space strategy. From those deliberations, three possible desired actions may be reached: decide to become stronger, keep the status quo, or become weaker.[63] The final decision and strategy chosen may be based on what course of action is in its most vital interest.

The first option is perhaps most readily understandable within the context of international affairs and comparative strategies. This option is appropriate when those with less power and influence seek to improve their situation and promote their interests. Such an improvement may be achieved through either military or non-military methods.

The second option is when it is in an emerging power's interest to maintain the status quo. Although it might be presumed that every emerging power should always want to improve its standing among space powers and increase its associated space capabilities, history illustrates that this is not necessarily true. The reason for this possibility is that there may be advantages to being an emerging space power. These advantages may occur when an emerging power is in a cooperative relationship or mutual defense agreement with another power that is able to exercise general and persistent command of space. Because of the benefits gained in such a cooperative relationship, it may be best to maintain the standing as an emerging space power. An emerging power can take advantage of a friendly and more capable power's technological developments in launch systems or satellites, without incurring large research and development costs itself. Furthermore, an emerging space power can maintain minimal space-related training and educational infrastructure, while still having considerable access to space by "piggy-backing" on the efforts of others. The monetary savings enjoyed in such a cooperative relationship can then be used for those non-space activities considered more critical. Therefore, a cooperative relationship of this kind enables a lesser space power to achieve many of same benefits as a more capable power or group of countries, without taking the same risks or expending the same amount of resources.

The third option is when an emerging power would want its influence and capabilities in space reduced. Although such instances may seem few, they do exist. These cases may occur when domestic or world economic conditions require cutting costs in space-related activities and research. When a temporary economic downturn is expected, and more pressing national security problems exist, a scaling back in space activities may be warranted. Such a short-term scaling back can be pursued, knowing that an increase in space activities is planned once economic conditions improve. Moreover, another instance when a reduced role in space is warranted is when it is desired to minimize associated risk or exposure to space-based threats.

A state actor, for instance, that has historically relied on space-based satellite communications for the transfer of news, data, and information might decide to increase its use of fiber optic cables or terrestrial-based cellular phone systems instead of using satellite communications. In such a case, the reliance on space-based communications is reduced, while a proportional increase in non-space-based communications offsets the difference. Using the parlance of the U.S. space community, such an action might be considered part of space mission assurance and resilience, because it employs alternative methods through diversification of capabilities within the other domains of operation.[64] A move such as this that intentionally reduces involvement in space activities may be seen as a suitable method of reducing vulnerability to future space-based attack.

Despite there being three possible courses of action, the remainder of this chapter will discuss strategies related to the first option—gaining more influence, becoming stronger, and contesting the command of another space power. This discussion is not intended to discount the other two options, but those topics are better suited for a more thorough treatment in a separate work.

Non-military actions

Emerging space powers will have non-military levers of power to advance national interests. By definition, a less capable space power will be more challenged in influencing those considered more capable; nonetheless, non-military actions can be used to achieve some strategic objectives. In fact, non-military approaches may be the most appropriate method to advance national or strategic objectives for those considered least capable, because situations where it is advantageous to seek military confrontation may be few. This section will address the benefits associated with diplomatic actions; promoting national pride; and benefits coming from a more technologically sophisticated polity.

Diplomatic initiatives

As with great and medium space powers, emerging powers may use diplomacy to contest the power and influence of another. Long-term diplomatic initiatives and activities are commensurate with a cumulative strategy, as described by J.C. Wylie.[65] That is because one success does not usually depend on the one previous, but the benefits coming from several successes, even if sporadic, can build upon each other for a cumulative gain.

One potential venue for emerging space powers to advance individual or collective interests is the Committee on the Peaceful Uses of Outer Space (COPUOS). The United Nations General Assembly established the permanent Committee in 1959, and since that time the number of members has grown from 24 to 84.[66] Using the previous definitions of great space powers and medium space powers, there are over 50 countries that could be considered as having influence regarding the use of space, even without having a significant space-related capability themselves.[67] This group of countries can seek to shape space discussions and perceptions through the COPUOS, especially given the Committee's responsibilities for reviewing international cooperation in the peaceful uses of outer space; studying space-related activities that could be undertaken by the United Nations; encouraging space research programs; and studying legal problems arising from the exploration of outer space.[68]

Canadian officials see the Committee's role as being an essential element in space governance to increase the socio-economic benefits of space.[69] As a result, Canada has used its position on COPUOS to advance issues that it sees in its interest, as well as that of the global community. This agenda includes issues related to maintaining a safe, predictable, and sustainable outer space environment; implementing transparency and confidence building measures in outer space activities; addressing challenges caused by space debris, space weather and emerging space activities; promoting diversity and women in space; and underscoring how space can improve global health.[70]

Saudi Arabia has been using diplomatic channels to advance its interests, which include pursuing deeper cooperation with Russia to share in space exploration efforts. This is exemplified by H.M. King Salman of Saudi Arabia signing an agreement with President Vladimir Putin of Russia in October 2017, committing both countries to space exploration cooperation. This Saudi–Russia cooperation extends to a variety of sectors, including space exploration, energy, and defense. This move by Saudi Arabia is thought to include the desire to persuade Russia to reign in Iranian influence through the use of economic and security inducements that Iran cannot possibly provide.[71]

Instilling national pride

For emerging space powers, an ambitious space program may not be an option, but they can use space-related activities to bolster national pride. This is indeed the case with Canada. The Canadarm-2 and Dextre—robotic arms and manipulators used to build and now maintain the ISS—are both featured on the Canadian five-dollar bill and symbolize Canada's ongoing contribution to the international space program.[72] Retired Canadian astronaut Chris Hadfield revealed the bill while serving as commander of the ISS. He noted that engineering and exploration can become something more, and said "It can become culturally symbolic of what people can achieve and part of what people identify as what they are proud of in their civilization."[73]

Additionally, when joining with another country's ambitious endeavor in space, an emerging space power can gain a sense of benefit coming from active participation. This benefit may include having one's own citizens fly aboard another country's spacecraft to instill national pride. For instance, the ISS is an international partnership of space agencies, with principal space agencies including those of the United States, Russia, member states of the European Space Agency, Japan, and Canada.[74] The space station demonstrates how states of various degrees of space capability and participation may all benefit from sharing in advanced space-based science and research. Other ways of bolstering national pride include activities like launching the first national satellite into orbit using third party or commercial launch services. This event may instill pride through a country being able to claim it is now a "space power." Low-cost, low-risk initiatives such as this are meant to increase the optimism of a populace and may result in more domestic support for those governmental leaders in power.

Modest achievements in space can be communicated through media outlets, whether through traditional news outlets or social media. By conducting a sustained campaign to promote news that advances its long-term strategy, an emerging space power may, over time and using a cumulative strategy, change what is perceived or considered as fact by others. Depending on the desired strategy, long-term advocacy through public affairs and strategic communications methods can advance national pride and standing amongst the international space community.

Technically educated workforce

In today's global economy, having and maintaining a competitive workforce is frequently considered a critical factor. Because of this view, some governments have sought to increase the percentage of their population who are educated and trained in science, technology, engineering, and math career fields. Such education and training are often deemed essential for a growing and robust domestic economy, especially one that is able to weather global economic downturns.

Emerging powers may take a lesson from the pages of the U.S. space program of the 1960s and 1970s. During this time, there was an implicit expectation among U.S. leaders to employ and train as large a number of people as possible in the process of getting a man to the Moon.[75] NASA built operations and support centers across the country and hired a technical workforce to support the national agenda. This expansion in aerospace industrial capability; associated infrastructure; and a science, technology, engineering, and math educated workforce had a cascading benefit enabling the commercial aerospace industry to also expand during this timeframe.

Emerging space powers may also use this strategy of developing a highly-educated, technology-savvy workforce that enables the growth of a burgeoning space industry. There is often a hope that a more educated workforce will lead to additional benefits, including gaining access to new markets, branching into new high-technology industries, increasing economic growth, and creating more technology-focused jobs.[76] The hope is such results will enhance a country's economic instrument of national power, which can potentially lead to improved effectiveness in diplomatic initiatives and basic research and development as well.

Recent developments in Australian space policy demonstrate this concept. In May 2018, as part of its annual budgetary process, the Australian Government announced the establishment of an Australian Space Agency with ongoing funding starting at $26 million over the next four years.[77] The stated objective of the new agency is to support the development of Australia's space industry so that it can compete effectively in the global space sector.[78] Creating the space agency is hoped to develop the country's space industry, while also opening the door to its official participation in larger international space science missions. A media announcement by the Australian government states the new space agency will help Australia's businesses win a greater share of the multi-billion dollar global space market and establish a new space industry, with the potential of creating 20,000 new jobs.[79] Another benefit of the space agency is thought to be to "inspire the nation, especially young kids," according to Anna Moore, an astronomer at Australian National University in Canberra.[80]

Fluidity between the offense and defense

> Take advantage of the enemy's unpreparedness; attack him when he does not expect it; avoid his strength and strike his emptiness, and like water, none can oppose you.
>
> Sun Tzu[81]

Unlike the previous discussions on great and medium space powers in Chapters 5 and 6 where offensive and defensive strategies were discussed separately—albeit interdependently—it is best to describe potential military actions of emerging powers as a constant tension between offensive and defensive strategies. This constant tension is out of necessity. An emerging power is not likely to win a decisive military victory against a superior power, but the less capable power can be soundly defeated in a single engagement if it does not take adequate precautions. Clausewitz warned of this. A persistent fear of insurgent fighters is a quick and decisive defeat, and consequently, such insurgent warfare "... must not be decided by a single stroke."[82] So, while offensive approaches are needed to advance political ends, achieve positive aims, and impact negatively the enemy's military, defensive approaches must be closely integrated to prevent defeat and ensure the long-term viability of a less capable power's cause or objectives.

Mao notes the importance of offensive actions in defeating the enemy, along with the interdependence of offensive and defensive methods in an overall strategy. This is reflected in Mao's quote at the beginning of the chapter on striving to preserve one's forces while destroying that of the enemy.[83] Furthermore, Mao goes on to observe:

> Flexibility in the employment of forces revolves around the effort to take the offensive, and planning likewise is necessary chiefly in order to ensure success in offensive operations. Measures of tactical defence are meaningless if they are divorced from their role of giving either direct or indirect support to an offensive. Quick decision refers to the tempo of an offensive, and exterior lines refer to its scope. The offensive is the only means of destroying the enemy and is also the principal means of self-preservation, while pure defence and retreat can play only a temporary and partial role in self-preservation and are quite useless for destroying the enemy.[84]

The context of Mao's comment was not that defensive approaches are unimportant, but defensive action must directly support offensive actions in achieving political ends and eventual success.

Because an emerging power will have limited military resources when compared to a stronger power, incorporating defensive strategies or measures may be the most effective and efficient means to protect space-related interests. A defensive intent attempts to prevent an opponent from accomplishing something or attempts to protect something that is already held. Based upon the general theory of Clausewitz, less force or energy is usually required, as opposed to that needed when taking something by force. It is worth underscoring again that taking a truly defensive posture includes making the necessary preparations, given the likely means and timing of an adversary's anticipated offensive actions. Not making adequate preparations, therefore, means a sound defensive strategy has not been incorporated and a decision—whether intentional or not—has been made to be remain vulnerable.

Mao emphasized the need for preparations in saying,

> Without planning, victories in guerrilla warfare are impossible. Any idea that guerrilla warfare can be conducted in haphazard fashion indicates either a flippant attitude or ignorance of guerrilla warfare. The operations in a guerrilla zone as a whole, or those of a guerrilla unit or formation, must be preceded by as thorough planning as possible, by preparation in advance for every action. Grasping the situation, setting the tasks, disposing the forces, giving military and political training, securing supplies, putting the equipment in good order, making proper use of the people's help, etc.—all these are part of the work of the guerrilla commanders, which they must carefully consider and conscientiously perform and check up on.[85]

When giving examples of the planning that must be done ahead for military engagements, Mao's list connotes the need to understand the situation, know the capabilities and vulnerabilities of one's forces and assets, and those measures that should be taken to improve the probability for success. This advice will hold for space strategy as well.

Thomas Schelling's idea of *deterrent defense* is also applicable when considering an emerging space power's integrated offensive and defensive strategy. Schelling believed that war can have either a deterrent or compellent intention, just as it can have defensive and offensive aims.[86] He writes that if the object is to induce the adversary not to proceed on some course of action, then one can make the enemy's encroachment painful or costly.[87] In writing about those less capable militarily, Schelling says, "Resistance that might otherwise seem futile can be worthwhile if, though incapable of blocking progress, it can nevertheless threaten to make the cost too high."[88] He calls this "dynamic" deterrence in which the threat is communicated by progressive fulfillment.[89] It is possible to deter an adversary's repeated action through defensive means—even though it has no hope of repelling enemy action—by making it too costly for any aggression to be considered "successful."[90] These types of considerations are in agreement with a "cost versus benefit" calculus for deterrence and military operations.

Based upon the writings of Mao and Schelling, it is improbable that minor actions by emerging powers will in themselves be able to decide the outcome of a war or conflict—at least when considered as isolated incidents. Nonetheless, these actions can still achieve modest results. Minor actions can prevent the superior power from increasing its command of space and cause it to expend more resources and personnel to counter the threat of attack. If it is perceived that the minor attack has been a success, an emerging power's domestic morale may improve.

The next sections will describe other relevant topics related to integrated offensive and defensive operations, and these topics are based upon writings on the general theory of war. While the context of this chapter is space strategy for emerging powers, the approaches described can be part of a multi-domain warfare or "all aspects unified" operations as described in Chinese strategy.[91] Also, the methods described may be part of terrestrial actions that seek to affect negatively the space activities of another.

Dispersal and concentration

> Move when it is advantageous and create changes in the situation by dispersal and concentration of force.
>
> Sun Tzu[92]

Mao's writing on dispersal and concentration of forces underpins much of his thinking on guerrilla warfare. He wrote, "Although the flexible dispersal or concentration of forces according to circumstances is the principal method in guerrilla warfare, we must also know how to shift (or transfer) our forces flexibly."[93] Yet Sun Tzu's preceding quote is perhaps clearer about the matter of dispersal and concentration. As illustrated, the constant movement between dispersal and concentration in war is a foundational element in the writings of both renowned strategists. As with fluidity between offensive and defensive operations, the use of dispersal and concentration will be a necessary centerpiece of an emerging power's military strategy. Offensive actions should concentrate against the adversary to achieve maximum effect, and when not conducting offensive actions, forces and assets should be dispersed to avoid detection and decisive defeat.

In writing about popular insurrections and the need for dispersal, Clausewitz observes that it "should be nebulous and elusive, its resistance should never materialize as a concrete body, otherwise the enemy can direct sufficient force at is core, crush it, and take many prisoners."[94] On how insurgencies should operate and disperse, he goes on to say that it is "... better to scatter and continue their resistance by means of surprise attacks, rather than huddle together in a narrow redoubt, locked into a regular defensive position from which there is no escape."[95]

When considering space strategy, one of the main reasons for dispersing forces is so that a superior force that exercises command of space is not able to defeat decisively a less capable space power during a single engagement. Dispersal can be achieved through land, sea, air, space, and cyber domains, which may facilitate space mission assurance and resiliency effectiveness.

The other part of the concept is concentration. The principle of concentration enables emerging powers to move resources where offensive operations are anticipated or where the potential threat of attack would be most damaging to the enemy. Concentration of forces or effects will be needed when conducting offensive action to achieve positive aims. By concentrating limited forces or effects within a region for a specific period, an emerging space power may be able to achieve a relative advantage over an adversary. This idea is in keeping with principles of local and temporary command of space, as discussed in Chapter 2.

Nibbling around the edges

Insurgency operations may operate around exterior lines of communication and should attempt to "nibble at the shell and around the edges" of the enemy's operating area, using the language of Clausewitz.[96] According to the Prussian strategist, these operations are meant to occur just outside where the adversary's strength lies, to deny him those areas altogether.[97] He comments that insurgency operations are aimed at gaining an ever increasing level of popular support, to the extent that "The flames will spread like a brush fire, until they reach the area on which the enemy is based, threatening his lines of communication and his very existence."[98] In the context of space strategy, by attacking along the periphery of celestial lines of communication (CLOCs) or in regions where the adversary is not strong, positive aims can be achieved in support of diplomatic, economic, informational, or military purposes. A less capable space force may attack an adversary's distant CLOCs, thereby avoiding a direct engagement where the preponderance of the enemy's forces and assets are located.[99]

From the lessons of Sun Tzu, Clausewitz, and Mao, it is expected that an emerging space power's strategy that seeks to employ force will include small-scale operations along exterior CLOCs, targeting easily accessible locations along these lines. Because of the technological sophistication required for operations in space, these types of operations may prove more difficult to implement and accomplish than those against terrestrial targets, at least in the near-term. Suitable insurgency strategies incorporating the idea of "nibbling around the edges" may entail:

- operations against terrestrial facilities used for uplinks or as central distribution hubs for space-based information;
- targeting an adversary's space agency headquarters, manufacturing facilities, and launch facilities because they are at the tail of space-related supply chains; and
- conducting non-kinetic operations—like through signal jamming and lasing—along the periphery of operations against an adversary's communication and electro-optical satellites in geostationary orbit.

Asymmetry and cyber actions

Sun Tzu believed in the need to adjust against a well-prepared and knowledge-able adversary, as exemplified by his statement, "Attack where he is unprepared;

sally out when he does not expect you."[100] This idea is incorporated in what is termed today as *asymmetric operations*, which seeks to circumvent an adversary's strengths and exploit its weakness through the use of dissimilar strategies, tactics, and capabilities.[101]

Emerging space powers may indeed include elements of asymmetric warfare. This inclusion is because of the need to achieve political objectives or impact negatively the enemy using limited military means. When considering asymmetric operations for emerging space powers, one action comes readily to mind: using one's strength in one domain of operation to affect the adversary negatively in its weaker domain of operation, such as ground-based operations against an adversary's launch facility, just prior to the adversary's launching a payload into orbit.

Moreover, when thinking about asymmetric operations and potential operational advantage, cyber actions can be a preferred method to achieve the desired effect. Because the cyber domain encompasses expansive lines of communication involving a global and multi-domain network—with servers, various network hardware, and terminals—cyberattacks may be seen as an easy method for those less capable to conduct asymmetric operations. Based upon current experience regarding the widespread use of cyber-attacks, this view seems to be true. Because the individual or group spearheading a cyber-attack—presuming it is not being conducted autonomously by computer software—can be located thousands of miles away from the area or locality to be impacted, any fear of adverse consequences from a cyber-attack can be seen as minimal.

Furthermore, it is often difficult to discern definitively through forensics and other intelligence and law enforcement methods who is responsible for many cyber intrusions and attacks. Without such forensics data and other corroborating information, an attribution decision that enables a retaliatory action from a cyber-attack of unknown origin seems doubtful, indeed. For these reasons, cyber-attacks or other adversary actions using the Internet would seem to be a viable means for a less capable power to avoid defeat and protect limited military resources, while seeking to exploit an adversary's weakness. These kinds of non-attributable and non-overt methods may be able to diminish the influence and effectiveness of the superior adversary, in the long-term and using a cumulative strategy.

Affecting strategic positions

> If you are able to hold critical points on his strategic roads the enemy cannot come.
>
> Sun Tzu[102]

Regarding the use of the military instrument of national power, emerging space powers will want to consider affecting strategic positions. Strategic positions may include high-value space assets or those considered of special importance. Affecting strategic positions is commensurate with Sun Tzu's quote above.[103] In his writings, Sun Tzu emphasized repeatedly the use of positions to gain strategic advantage, as further exemplified by his advice, "Therefore the skillful commander takes up a position in which he cannot be defeated and misses no opportunity to master his enemy."[104] By exploiting the advantages of strategic positions, the most reward for least risk is potentially achieved. Also, the destruction of an adversary's high-value space assets may have an economic effect because of its replacement cost and the effect on space-enabled business; consequently, there may be psychological repercussions following the loss of the high-value asset. The ability of insurgent warfare to cause psychological effects is noted by Clausewitz, who observes regarding the utility of guerrilla warfare, "… the psychological element, is called into being only by this type of usage."[105]

There are several methods that an emerging space power may use to exploit strategic positions, including:

- irreversible actions affecting—through either kinetic or non-kinetic means—a satellite providing specialized services;
- conducting operations against a ground station through which a high-volume of satellite communication passes;
- affecting negatively the manufacturer's facilities used in producing high-value satellites;
- conducting land, sea, air, space, or cyber operations against an adversary's land-based launch site, and in doing so potentially limit an adversary's ability to reconstitute space systems damaged during a conflict;
- affecting a space station that is a hub for scientific, commercial, logistical, or military enterprise in hopes of curtailing specialized activities and causing serious political repercussions; and
- affecting negatively permanent stations on the Moon or on other celestial bodies, because they will likely provide unique services or capabilities.

Protracted war

> For there has never been a protracted war from which a country has benefited.
>
> Sun Tzu[106]

The context of Sun Tzu's statement above is to highlight that a powerful, centralized state can be worn down through a protracted conflict. Yet, this style of warfare is likely to be advantageous to a less capable power waging war against the more powerful. While emerging powers—including insurgents—fight from a point of initial weakness compared to a state supported army, they can use the element of time against the state through the use of protracted conflict.

Because of the relative strength between belligerents, guerrilla warfare frequently calls for attacking a foe and then retreating before any substantial counter-attack by the stronger enemy can take place. Within his context of popular uprisings in land warfare, Clausewitz thought that because insurgents usually operate within the interior of their territory, a guerrilla strategy "calls for avoiding defeat by yielding the contested ground in time."[107] Clausewitz compared insurgencies to slow burning fires in writing:

> Like smoldering embers, it consumes the basic foundation of the enemy forces. Since it needs time to be effective, a state of tension will develop while the two elements interact. This tension will either gradually relax, if the insurgency is suppressed in some places and slowly burns itself out in others, or else it will build up to a crisis: a general conflagration closes in on the enemy, driving him out of the country before he is faced with total destruction.[108]

The purpose of protracting a conflict is to protect one's interests and prevent the enemy from accomplishing its aim until the situation changes in one's favor. Therefore, buying time through protracted warfare incorporates elements of defensive strategy. Moreover, minor actions can delay defeat, until allies or other forces can join the fight against the superior force.

When seeking to protract a war until the situation changes in one's favor, the concept includes using either non-military or military actions against the adversary. Examples of actions that may facilitate protracting a war include:

- military actions consistent with a cumulative strategy and the indirect approach;
- collective action with other states and space powers to enforce an economic embargo against an adversary in the hopes of affecting its space-reliant commerce;
- introducing legal actions into domestic or international courts to potentially delay an adversary's ability to pursue its immediate objectives, or what has been called *lawfare*; and
- releasing derisive media reports about an adversary's space activities with the intent of forcing an adversary to respond before resuming a contentious action.

Actions of terrorists

A sub-category of insurgent warfare is terrorism. Ken Pollack observes, "Terrorism, of course, is a form of insurgent warfare. Although not all insurgencies employ terrorism, most have historically done so, and every terrorist group is, by definition, employing a form of insurgent warfare."[109] As with insurgents, the goals of terrorists may be political in nature, yet some terrorists may not have purely political goals. Instead, terrorists may want anarchy, chaos, or the disestablishment of a state. Terrorist organizations may use any of the general strategies relevant to emerging powers. Yet unlike most emerging powers, the aims of terrorists will not only be to contest the command of another, but also may be to cause fear and mass casualties. Because of these aims, terrorists will likely prefer to attack easily accessible locations that cause the most sensational reactions from the general population.

J.C. Wylie and Thomas Schelling both write on the strategy of terrorists. Wylie considered the general strategy of terrorism under his general theory of control and writes,

> Accordingly, terrorists aim is for some selected degree of control of the processes of social change for their own purpose. Terrorists seek to achieve this by control of the pattern of their war against society. Terrorists do this by creating and manipulating a center of gravity (either a person or an installation that will ensure public attention) that they have selected to the advantage of the strategist and the disadvantage of the organized society over which terrorists want to exercise control.[110]

Considering the actions of terrorists, Schelling saw terrorism as violence intended to coerce the enemy rather than to weaken him militarily.[111] In discussing terrorists needing to appreciate the political value of the potential use of violence compared to whatever value they may attach to solely destructive actions, Thomas Schelling writes,

> Smart terrorists—and the people who might assemble nuclear explosive devices, if they can get the fissionable material, will have to be highly intelligent—should be able to appreciate that such weapons have a comparative advantage toward *influence*, not simple destruction.[112]

Overall, both Wylie and Schelling share similar thinking. For Wylie, the terrorists' aim is control of the pattern of conflict; for Schelling, the aim is to coerce the enemy. Therefore, both understood that terrorists seek to influence the actions and thinking of others.

Some potential space-related targets that terrorists may consider acting against include:

- corporate headquarters involved in the development or use of space systems;
- ground-based relay stations that support space-based commerce and trade because they are numerous and may not be well protected;
- manned spacecraft readying for launch because the destruction will cause sensational reactions and result in media attention;
- high-value commercial space systems, such as those providing one-of-a-kind services or with high brand recognition; and
- manned space stations to cause catastrophic damage (admittedly, attacking manned space stations in orbit or cislunar space presently may be the most attractive option, yet technologically challenging for terrorists to act against).

Conclusion

Even though the strategies for emerging powers are less understood than the strategies of those considered more powerful, these strategies are important and should be studied. The specific space strategies for emerging space powers are likewise important. There are more emerging space powers than great or medium space powers; therefore, the strategies of less capable space powers will be an enduring consideration for space strategists.

Based upon the writings of Sun Tzu, Clausewitz, Mao, Liddle Hart, and Wylie, the concepts of cumulative strategy and the indirect approach will often be most appropriate for emerging space powers. As with the strategy of insurgents and consistent with the teachings of Clausewitz and Mao, emerging space powers will want to avoid any situation that may result in a decisive defeat against one's forces. This condition may require employing operations along the periphery of the adversary's CLOCs. A sound strategy will require ample preparations and defensive measures, to protect forces and assets until offensive actions are appropriate. Ultimately, offensive actions will be important to achieve positive aims and victory, yet these actions must be integrated into defensive strategy.

Notes

1 Mao Tse-tung, *Selected Military Writings of Mao Tse-tung* (Seattle, WA: Praetorian-press.com, 2011) Kindle edition, location 3570.
2 Deganit Paikowsky, "The Space Club—Space Policies and Politics" (paper presented at the 60th International Astronautical Congress, Daejeon, Republic of Korea, October 2009). Also referenced by Scott Pace, "A Space Launch without a Space Program," *38 North*, April 11, 2012, http://38north.org/2012/04/space041112
3 Dave Baiocchi and William Welser IV, "The Democratization of Space: New Actors Need New Rules," *Foreign Affairs* (May/June 2015), www.foreignaffairs.com/articles/space/2015–04–20/democratization-space
4 Ibid.
5 The White House, *The National Security Strategy of the United States of America* (December 2017), 31, www.whitehouse.gov/wp-content/uploads/2017/12/NSS-Final-12–18–2017–0905.pdf
6 Sandra Erwin, "Space Industry takes Prominent Role in Trump's National Security Strategy," *Spacenews*, December 18, 2017, http://spacenews.com/space-industry-takes-prominent-role-in-trumps-national-security-strategy/
7 The White House, *The National Security Strategy of the United States of America* (December 2017), 31.
8 Carl von Clausewitz, *On War*, trans. and eds. Michael Howard and Peter Paret (Princeton, NJ: Princeton University Press, 1989), 479–483.
9 Ibid.
10 Ibid., 483.
11 Ibid., 479.
12 John Palimaka, "The 30th Anniversary of Alouette I," *IEEE Canadian Review* (Fall 1992), www.ieee.ca/millennium/alouette/alouette_impact.html
13 Palimaka, "The 30th Anniversary of Alouette I."
14 "Alouette I and II," Canadian Space Agency, last modified March 5, 2012, www.asc-csa.gc.ca/eng/satellites/alouette.asp

15　"Black Brant Sounding Rocket," Royal Aviation Museum of Western Canada, accessed August 27, 2018, www.royalaviationmuseum.com/759/blackbrant/

16　Lydia Dotto, *Canada and The European Space Agency Three Decades of Cooperation*, European Space Agency (2002), www.esa.int/esapub/hsr/HSR_25.pdf

17　"Canadarm and Canadarm2 – A comparative table," Canadian Space Agency, last modified March 22, 2018, www.asc-csa.gc.ca/eng/iss/canadarm2/canadarm-canadarm2-comparative-table.asp

18　Ibid.

19　"Curiosity and the Mars Science Laboratory Mission," Canadian Space Agency, last modified May 5, 2017, www.asc-csa.gc.ca/eng/astronomy/mars/curiosity.asp

20　"James Webb Telescope Overview," National Aeronautics and Space Administration, accessed August 27, 2018, www.nasa.gov/mission_pages/webb/about/index.html

21　"Space Missions," Canadian Space Agency, last modified April 27, 2018, www.asc-csa.gc.ca/eng/missions/default.asp

22　"Biography of Marc Garneau," Canadian Space Agency, last modified August 2006, www.asc-csa.gc.ca/eng/astronauts/canadian/former/bio-marc-garneau.asp

23　Allan Woods, "Chris Hadfield: The Superstar Astronaut taking Social Media by Storm," *Guardian*, February 22, 2013, www.theguardian.com/science/2013/feb/22/chris-hadfield-canada-superstar-astronaut

24　"Space Situational Awareness and the Sapphire Satellite," National Defence and the Canadian Armed Forces, last modified January 30, 2014, www.forces.gc.ca/en/news/article.page?doc=space-situational-awareness-and-the-sapphire-satellite/hr0e3oag

25　Ibid.

26　William E. Burrows, *This New Ocean: The Story of the First Space Age* (New York: Random House Inc, 1998), 621.

27　"About," Arabsat, accessed July 28, 2018, www.arabsat.com/english/about

28　"Arabsat 1A, 1B, 1C/Insat 2DT," Gunter's Space Page, accessed July 28, 2018, http://space.skyrocket.de/doc_sdat/arabsat-1a.htm

29　Burrows, *This New Ocean*, 553.

30　Rym Ghazal, "The First Arab in Space," *The National*, April 9, 2015, www.thenational.ae/arts-culture/the-first-arab-in-space-1.32633

31　Caleb Henry, "Arabsat Falcon Heavy Mission Slated for December-January Timeframe," *Spacenews*, June 1, 2018, https://spacenews.com/arabsat-falcon-heavy-mission-slated-for-december-january-timeframe/

32　"Saudi Arabia and Earth Observation Systems," GlobalSecurity.org, accessed July 28, 2018, www.globalsecurity.org/space/world/saudi/intro.htm

33　"SaudiComsat 1, 2, 3, 4, 5, 6, 7," Gunter's Space Page, accessed July 28, 2018, http://space.skyrocket.de/doc_sdat/saudicomsat-1.htm; "SaudiSat 3," Gunter's Space Page, accessed July 28, 2018, http://space.skyrocket.de/doc_sdat/saudisat-3.htm

34　"SaudiSat 4," Gunter's Space Page, accessed July 28, 2018, http://space.skyrocket.de/doc_sdat/saudisat-4.htm

35　"Space and Aeronautics," King Abdulaziz City for Science and Technology, accessed July 28, 2018, www.kacst.edu.sa/eng/rd/pages/content.aspx?dID=97

36　"France-Saudi Arabia Space Cooperation, CNES and KACST Sign Executive Program Agreement," CNES, press release, April 10, 2018, https://presse.cnes.fr/en/france-saudi-arabia-space-cooperation-cnes-and-kacst-sign-executive-programme-agreement; John Sheldon, "Saudi Arabia and Russia Deepen Space Cooperation, Agree on Joint Space Exploration Projects," *Spacewatch.Global*, October 9, 2017, https://spacewatchme.com/2017/10/saudi-arabia-russia-deepen-space-cooperation-agree-joint-space-exploration-projects/

37　Andrew Jones, "Chang'e-4: Far Side of the Moon Lander and Rover Mission to Launch in December," *gbtimes*, June 18, 2018, https://gbtimes.com/change-4-far-side-of-the-moon-lander-and-rover-mission-to-launch-in-december

38 MacDonald Dettwiler and Associates acquired DigitalGlobe in October 2017. Andrea Shalal, "DigitalGlobe Forms Satellite Joint Venture with Saudi Firms," *Reuters*, February 21, 2016, www.reuters.com/article/us-digitalglobe-saudi-venture/digitalglobe-forms-satellite-joint-venture-with-saudi-firms-idUSKCN0VU11F

39 Peter B. de Selding, "DigitalGlobe and Saudi Government Sign Joint Venture on Satellite Imaging Constellation," *Spacenews*, February 22, 2016, https://spacenews.com/digitalglobe-and-saudi-government-sign-joint-venture-on-satellite-imaging-constellation/

40 Jeff Foust, "Virgin Signs Agreement with Saudi Arabia for Billion-Dollar Investment," *Spacenews*, October 26, 2017, https://spacenews.com/virgin-signs-agreement-with-saudi-arabia-for-billion-dollar-investment/; John Sheldon, "Saudi Arabia Rumored to be Funding Ukrainian Hypersonic Spaceplane," *Spacewatch. global*, March 28, 2018, https://spacewatch.global/2018/03/saudi-arabia-rumoured-funding-ukrainian-hypersonic-spaceplane/

41 "Saudi Arabia's Vision 2030 Plan Is Too Big to Fail – Or Succeed," *Stratfor*, July 27, 2018, https://worldview.stratfor.com/article/saudi-arabias-vision-2030-plan-too-big-fail-or-succeed

42 John B. Sheldon, "Saudi Arabia's Vision 2030: A Golden Opportunity for Space?" ThorWatch Analysis (ThorGroup GmbH, May 1, 2016), 4.

43 Department of Defense, *2018 National Defense Strategy of the United States of America: Sharpening the American Military's Competitive Edge* (2018), 3, www.defense.gov/Portals/1/Documents/pubs/2018-National-Defense-Strategy-Summary.pdf

44 Department of Defense, *2018 National Defense Strategy of the United States of America*, 3. Emphasis original.

45 Alan Boyle, "Interview: Jeff Bezos Lays Out Blue Origin's Space Vision, from Tourism to Off-planet Heavy Industry," *Geek Wire*, April 13, 2016, www.geekwire.com/2016/interview-jeff-bezos/; Dave Mosher, "Elon Musk Has Published a New Study About His Ambitious Plans to Colonize Mars with SpaceX," *Business Insider*, March 27, 2018, www.businessinsider.com/elon-musk-mars-colony-details-new-space-study-2018–3

46 "About us," Intelsat, accessed August 27, 2018, www.intelsat.com/about-us./

47 "About us," Inmarsat, accessed August 27, 2018, www.inmarsat.com/about-us/

48 J.C. Wylie, *Military Strategy: A General Theory of Power Control*, with introduction by John B. Hattendorf (New Brunswick, NJ: Rutgers University Press, 1967; reprint, Annapolis, MD: Naval Institute Press, 1989), 22–27.

49 Ibid., 24.

50 Ibid., 25.

51 Ibid., 22–27.

52 Ibid., 25.

53 Ibid., 48–49.

54 Ibid.

55 Ibid., 54.

56 Ibid., 48–49.

57 Ibid., 55.

58 Ibid., 60.

59 B.H. Liddell Hart, *Strategy: The Indirect Approach*, 2nd ed. (London: Faber and Faber, 1967), 337.

60 Mao Tse-tung, *Selected Military Writings*, location 5155.

61 Ibid., location 3727.

62 Ibid., location 5184.

63 The three possible choices are germane to medium space powers as well.

64 Office of the Assistant Secretary of Defense for Homeland Defense and Global Security, *Space Domain Mission Assurance: A Resilience Taxonomy* (September

2015), 6–7, http://policy.defense.gov/Portals/11/Space%20Policy/ResilienceTaxonomy WhitePaperFinal.pdf?ver=2016-12-27-131828-623

65 Wylie, *Military Strategy*, 22–27.
66 "Members of the Committee on the Peaceful Uses of Outer Space," United Nations Office for Outer Space Activities, accessed August 27, 2018, www.unoosa.org/oosa/en/members/index.html
67 The European Space Agency (ESA) has 22 Member States. See "What is ESA?" European Space Agency, accessed August 27, 2018, www.esa.int/About_Us/Welcome_to_ESA/What_is_ESA
68 "Committee on the Peaceful Uses of Outer Space," United Nations Office for Outer Space Activities, accessed August 27, 2018, www.unoosa.org/oosa/en/ourwork/copuos/index.html
69 Sylvain Laporte, President of the Canadian Space Agency, Statement to the Committee on the Peaceful Uses of Outer Space, UNISPACE+50 High-Level Segment, Vienna, Austria, June 20–21, 2018, 5, www.unoosa.org/documents/pdf/copuos/2018/hls/04_05EF.pdf
70 Laporte, President Canadian Space Agency, Statement to the Committee on the Peaceful Uses of Outer Space, 6.
71 Sheldon, "Saudi Arabia and Russia Deepen Space Cooperation."
72 Robert Z. Pearlman, "Canada Launches New Space Robot-Themed $5 Bill into Circulation," *Space*, November 7, 2013, www.space.com/23511-canada-launches-space-five-dollar-bill.html
73 Pearlman, "Canada Launches New Space Robot-Themed $5 Bill into Circulation."
74 "International Cooperation," National Aeronautics and Space Administration, accessed August 27, 2018, www.nasa.gov/mission_pages/station/cooperation/index.html
75 Joan Johnson-Freese, "China's Manned Space Program," *Naval War College Review* vol. 56 no. 3 (2003), 54.
76 The White House, *National Space Transportation Policy* (November 21, 2013), 1, https://obamawhitehouse.archives.gov/sites/default/files/microsites/ostp/national_space_transportation_policy_11212013.pdf
77 "Australian Space Agency," Australian Government Department of Industry, Innovation and Science, accessed August 27, 2018, https://industry.gov.au/INDUSTRY/IndustrySectors/SPACE/Pages/default.aspx
78 Ibid.
79 "Turnbull Government Launches Australia's First Space Agency," Australian Government Department of Industry, Innovation and Science, press release, May 14, 2018, www.minister.industry.gov.au/ministers/cash/media-releases/turnbull-government-launches-australias-first-space-agency
80 Dennis Normile, "Updated: Australia Creates Nation's First Space Agency," *Science*, May 8, 2018, www.sciencemag.org/news/2018/05/updated-australia-creates-nation-s-first-space-agency
81 Sun Tzu, *The Art of War* (*c*.400–320 BC), 4.20.
82 Clausewitz, *On War*, 480.
83 Mao Tse-tung, *Selected Military Writings*, location 3570.
84 Ibid., location 3749.
85 Ibid., location 3734.
86 Thomas C. Schelling, *Arms and Influence* (New Haven, CT: Yale University Press, 1966), 80.
87 Ibid., 78–79.
88 Ibid., 79.
89 Ibid.
90 Ibid.
91 Dean Cheng, *Cyber Dragon: Inside China's Information Warfare and Cyber Operations* (Santa Barbara, CA: Praeger Security International, 2017), 162.

92   Sun Tzu, *The Art of War* (*c.*400–320 BC), 7.12.
93   Mao Tse-tung, *Selected Military Writings*, location 3718.
94   Clausewitz, *On War*, 481.
95   Ibid., 482.
96   Ibid., 480–481.
97   Ibid.
98   Ibid., 481.
99   This is in keeping with Clausewitz's idea to "nibble at the shell and around the edges." Clausewitz, "The People in Arms," in *On War*, 479–483.
100 Sun Tzu, *The Art of War*, trans. Samuel B. Griffith (Oxford: Oxford University Press, 1963), 69.
101 Joint Chiefs of Staff, *Department of Defense Dictionary of Military and Associated Terms*, Joint Publication 1–02 (November 8, 2010, amended through February 15, 2016), 17, https://fas.org/irp/doddir/dod/jp1_02.pdf
102 Sun Tzu, *The Art of War* (*c.*400–320 BC), 6.3.
103 Sun Tzu, *The Art of War*, 96. Attributed to Tu Yu. Passage goes on to say, "When a cat is at the rat hole, ten thousand rats dare not come out; when a tiger guards the ford, ten thousand deer cannot cross."
104 Ibid., 87.
105 Clausewitz, *On War*, 479.
106 Sun Tzu, *The Art of War* (*c.*400–320 BC), 2.7.
107 Clausewitz, *On War*, 469.
108 Ibid., 480.
109 Kenneth Pollack, *A Path out of the Desert: A Grand Strategy for America and the Middle East* (New York: Random House, 2009), 188.
110 Wylie, *Military Strategy*, 105–106.
111 Schelling, *Arms and Influence*, 17.
112 Ibid., ix.

Bibliography

"About." Arabsat. Accessed July 28, 2018. www.arabsat.com/english/about

"About us." Inmarsat. Accessed August 27, 2018. www.inmarsat.com/about-us/

"About us." Intelsat. Accessed August 27, 2018. www.intelsat.com/about-us./

"Alouette I and II." Canadian Space Agency. Last modified March 5, 2012. www.asc-csa. gc.ca/eng/satellites/alouette.asp

"Arabsat 1A, 1B, 1C/Insat 2DT." Gunter's Space Page. Accessed July 28, 2018. http:// space.skyrocket.de/doc_sdat/arabsat-1a.htm

"Australian Space Agency" Australian Government Department of Industry, Innovation and Science. Accessed August 27, 2018. https://industry.gov.au/INDUSTRY/Industry Sectors/SPACE/Pages/default.aspx

Baiocchi, Dave and William Welser IV. "The Democratization of Space: New Actors Need New Rules." *Foreign Affairs* (May/June 2015). www.foreignaffairs.com/articles/ space/2015-04-20/democratization-space

"Biography of Marc Garneau." Canadian Space Agency. Last modified August 18, 2006. www.asc-csa.gc.ca/eng/astronauts/canadian/former/bio-marc-garneau.asp

"Black Brant Sounding Rocket." Royal Aviation Museum of Western Canada. Accessed August 27, 2018. www.royalaviationmuseum.com/759/blackbrant/

Boyle, Alan. "Interview: Jeff Bezos Lays Out Blue Origin's Space Vision, from Tourism to Off-planet Heavy Industry." *Geek Wire*. April 13, 2016. www.geekwire.com/2016/ interview-jeff-bezos/

Burrows, William E. *This New Ocean: The Story of the First Space Age.* New York: Random House Inc., 1998.

"Canadarm and Canadarm2 – A comparative table." Canadian Space Agency. Last modified March 22, 2018. www.asc-csa.gc.ca/eng/iss/canadarm2/canadarm-canadarm2-comparative-table.asp

Cheng, Dean. *Cyber Dragon: Inside China's Information Warfare and Cyber Operations.* Santa Barbara, CA: Praeger Security International, 2017.

Clausewitz, Carl von. *On War.* Translated and edited by Michael Howard and Peter Paret. Princeton, NJ: Princeton University Press, 1989.

"Committee on the Peaceful Uses of Outer Space." United Nations Office for Outer Space Activities. Accessed August 27, 2018. www.unoosa.org/oosa/en/ourwork/copuos/index.html

"Curiosity and the Mars Science Laboratory Mission." Canadian Space Agency. Last modified May 5, 2017. www.asc-csa.gc.ca/eng/astronomy/mars/curiosity.asp

Department of Defense. *2018 National Defense Strategy of the United States of America: Sharpening the American Military's Competitive Edge.* 2018. www.defense.gov/Portals/1/Documents/pubs/2018-National-Defense-Strategy-Summary.pdf

Dotto, Lydia. *Canada and the European Space Agency Three Decades of Cooperation.* European Space Agency. 2002. www.esa.int/esapub/hsr/HSR_25.pdf

Erwin, Sandra. "Space Industry takes Prominent Role in Trump's National Security Strategy." *Spacenews.* December 18, 2017. http://spacenews.com/space-industry-takes-prominent-role-in-trumps-national-security-strategy/

Foust, Jeff. "Virgin Signs Agreement with Saudi Arabia for Billion-Dollar Investment." *Spacenews.* October 26, 2017. https://spacenews.com/virgin-signs-agreement-with-saudi-arabia-for-billion-dollar-investment/

"France–Saudi Arabia Space Cooperation, CNES and KACST Sign Executive Program Agreement." CNES. Press release, April 10, 2018. https://presse.cnes.fr/en/france-saudi-arabia-space-cooperation-cnes-and-kacst-sign-executive-programme-agreement

Ghazal, Rym. "The First Arab in Space." *The National.* April 9, 2015. www.thenational.ae/arts-culture/the-first-arab-in-space-1.32633

Henry, Caleb. "Arabsat Falcon Heavy Mission Slated for December-January Timeframe." *Spacenews.* June 1, 2018. https://spacenews.com/arabsat-falcon-heavy-mission-slated-for-december-january-timeframe/

"International Cooperation." National Aeronautics and Space Administration. Accessed August 27, 2018. www.nasa.gov/mission_pages/station/cooperation/index.html

"James Webb Telescope Overview." National Aeronautics and Space Administration. Accessed August 27, 2018. www.nasa.gov/mission_pages/webb/about/index.html

Johnson-Freese, Joan. "China's Manned Space Program." *Naval War College Review* vol. 56 no. 3 (2003): 50–71.

Joint Chiefs of Staff. *Department of Defense Dictionary of Military and Associated Terms.* Joint Publication 1–02. November 8, 2010, amended through February 15, 2016. https://fas.org/irp/doddir/dod/jp1_02.pdf

Jones, Andrew. "Chang'e-4: Far Side of the Moon Lander and Rover Mission to Launch in December." *gbtimes.* June 18, 2018. https://gbtimes.com/change-4-far-side-of-the-moon-lander-and-rover-mission-to-launch-in-december

Laporte, Sylvain. President of the Canadian Space Agency. Statement to the Committee on the Peaceful Uses of Outer Space, UNISPACE+50 High-Level Segment. Vienna, Austria. June 20–21, 2018. www.unoosa.org/documents/pdf/copuos/2018/hls/04_05EF.pdf

Liddell Hart, B.H. *Strategy: The Indirect Approach*. 2nd edition. London: Faber and Faber, 1967.

Mao Tse-tung. *Selected Military Writings of Mao Tse-tung*. Seattle, WA: Praetorianpress.com, 2011. Kindle edition.

"Members of the Committee on the Peaceful Uses of Outer Space." United Nations Office for Outer Space Activities. Accessed August 27, 2018. www.unoosa.org/oosa/en/members/index.html.

Mosher, Dave. "Elon Musk Has Published a New Study About His Ambitious Plans to Colonize Mars with SpaceX." *Business Insider*. March 27, 2018. www.businessinsider.com/elon-musk-mars-colony-details-new-space-study-2018-3

Normile, Dennis. "Updated: Australia Creates Nation's First Space Agency." *Science*. May 8, 2018. www.sciencemag.org/news/2018/05/updated-australia-creates-nation-s-first-space-agency

Office of the Assistant Secretary of Defense for Homeland Defense and Global Security. *Space Domain Mission Assurance: A Resilience Taxonomy*. September 2015. http://policy.defense.gov/Portals/11/Space%20Policy/ResilienceTaxonomyWhitePaperFinal.pdf?ver=2016-12-27-131828-623

Pace, Scott. "A Space Launch without a Space Program." *38 North*. April 11, 2012. http://38north.org/2012/04/space041112

Paikowsky, Deganit. "The Space Club—Space Policies and Politics." Paper presented at the 60th International Astronautical Congress. Daejeon, Republic of Korea, October 2009.

Palimaka, John. "The 30th Anniversary of Alouette I." *IEEE Canadian Review* (Fall 1992). www.ieee.ca/millennium/alouette/alouette_impact.html

Pearlman, Robert Z. "Canada Launches New Space Robot-Themed $5 Bill into Circulation." *Space*. November 7, 2013. www.space.com/23511-canada-launches-space-five-dollar-bill.html

Pollack, Kenneth. *A Path out of the Desert: A Grand Strategy for America and the Middle East*. New York: Random House, 2008.

"Saudi Arabia and Earth Observation Systems." GlobalSecurity.org. Accessed July 28, 2018. www.globalsecurity.org/space/world/saudi/intro.htm

"Saudi Arabia's Vision 2030 Plan Is Too Big to Fail – Or Succeed." *Stratfor*. July 27, 2018. https://worldview.stratfor.com/article/saudi-arabias-vision-2030-plan-too-big-fail-or-succeed

"SaudiComsat 1, 2, 3, 4, 5, 6, 7." Gunter's Space Page. Accessed July 28, 2018. http://space.skyrocket.de/doc_sdat/saudicomsat-1.htm

"SaudiSat 3." Gunter's Space Page. Accessed July 28, 2018, http://space.skyrocket.de/doc_sdat/saudisat-3.htm

"SaudiSat 4." Gunter's Space Page. Accessed July 28, 2018. http://space.skyrocket.de/doc_sdat/saudisat-4.htm

Schelling, Thomas C. *Arms and Influence*. New Haven, CT: Yale University Press, 1966.

de Selding, Peter B. "DigitalGlobe and Saudi Government Sign Joint Venture on Satellite Imaging Constellation." *Spacenews*. February 22, 2016. https://spacenews.com/digitalglobe-and-saudi-government-sign-joint-venture-on-satellite-imaging-constellation/

Shalal, Andrea. "DigitalGlobe Forms Satellite Joint Venture with Saudi Firms." *Reuters*. February 21, 2016. www.reuters.com/article/us-digitalglobe-saudi-venture/digitalglobe-forms-satellite-joint-venture-with-saudi-firms-idUSKCN0VU11F

Sheldon, John. "Saudi Arabia Rumored to be Funding Ukrainian Hypersonic Spaceplane." *Spacewatch.global*. March 28, 2018. https://spacewatch.global/2018/03/saudi-arabia-rumoured-funding-ukrainian-hypersonic-spaceplane/

Sheldon, John. "Saudi Arabia and Russia Deepen Space Cooperation, Agree on Joint Space Exploration Projects." *Spacewatch.Global.* October 9, 2017. https://spacewatchme.com/2017/10/saudi-arabia-russia-deepen-space-cooperation-agree-joint-space-exploration-projects/

Sheldon, John. "Saudi Arabia's Vision 2030: A Golden Opportunity for Space?" Thor-Watch Analysis. ThorGroup GmbH, May 1, 2016.

"Space and Aeronautics." King Abdulaziz City for Science and Technology. Accessed July 28, 2018. www.kacst.edu.sa/eng/rd/pages/content.aspx?dID=97

"Space Missions." Canadian Space Agency. Last modified April 27, 2018. www.asc-csa.gc.ca/eng/missions/default.asp

"Space Situational Awareness and the Sapphire Satellite." National Defence and the Canadian Armed Forces. Last modified January 30, 2014. www.forces.gc.ca/en/news/article.page?doc=space-situational-awareness-and-the-sapphire-satellite/hr0e3oag

Sun Tzu. *The Art of War.* Translated by Samuel B. Griffith. Oxford: Oxford University Press, 1963.

Sun Tzu. *The Art of War. c.*400–320 BC

The White House. *National Security Strategy of the United States of America.* December 2017. www.whitehouse.gov/wp-content/uploads/2017/12/NSS-Final-12-18-2017-0905.pdf

The White House. *National Space Transportation Policy.* November 21, 2013. https://obamawhitehouse.archives.gov/sites/default/files/microsites/ostp/national_space_transportation_policy_11212013.pdf

"Turnbull Government Launches Australia's First Space Agency." Australian Government Department of Industry, Innovation and Science. Press release, May 14, 2018. www.minister.industry.gov.au/ministers/cash/media-releases/turnbull-government-launches-australias-first-space-agency

"What is ESA?" European Space Agency. Accessed August 27, 2018. www.esa.int/About_Us/Welcome_to_ESA/What_is_ESA

Woods, Allan. "Chris Hadfield: The Superstar Astronaut taking Social Media by Storm." *Guardian.* February 22, 2013. www.theguardian.com/science/2013/feb/22/chris-hadfield-canada-superstar-astronaut

Wylie, J.C. *Military Strategy: A General Theory of Power Control.* With introduction by John B. Hattendorf. New Brunswick, NJ: Rutgers University Press, 1967; reprint, Annapolis, MD: Naval Institute Press, 1989.

8 Space as a business domain

War is a matter not so much of arms as of money.

Thucydides[1]

As Thucydides' quote above illustrates, commerce, trade, and business have been inextricably linked with strategy for millennia. This linkage is because the economic instrument of national power will affect the available means used in conflict, along with potentially shaping the political ends sought. Economies help fund the weapons of war, and the desire for greater economic power can be viewed as a national interest that needs to be protected and advanced.

The commercial sector, particularly the technology and innovation that it enables, is recognized as critical to war, as illustrated in the 2018 U.S. National Defense Strategy, which states, "New commercial technology will change society and, ultimately, the character of war."[2] The commercial space sector is heavily influenced by emerging technologies and capabilities, and in this respect, commercial capabilities and services will affect the means and methods used in conflict. As a result, the commercial sector will influence the evolving character of warfare. While the technologies and companies referenced in this chapter will change, it is expected that the fundamental role of the commercial space sector will endure.

Based upon historical experience thus far, it is expected that economic and commercial activities in space will grow. This will make space-enabled business, commerce, and trade a national and global interest that should be protected. Highlighting this point, the 2017 U.S. National Security Strategy states, "As the U.S. Government partners with U.S. commercial space capabilities to improve the resilience of our space architecture, we will also consider extending national security protections to our private sector partners as needed."[3] Space strategy will need to incorporate and address this fact. Furthermore, the ever-increasing expansion of some commercial space services will make many functions nearly ubiquitous, thereby changing the character of warfare and our understanding of deterrence by denial in space strategy.

The space strategist must fully consider and integrate those space activities involved in commerce, trade, and business. Not integrating the commercial space industry will lead to unsound space strategy. Two areas that are little understood at present is how critical a part that commercial companies play in space strategy's purpose and in influencing space deterrence. This chapter will address those two areas, along with discussing the role of commercial space investors and emerging capabilities and services.

Promoting peace and stability

Commercial space companies provide a stabilizing influence on the international community. Commercial companies will often seek the free and open use of space, albeit while seeking competitive advantage within their market. Commercial satellite companies and service providers operate within the international legal regime, de-conflict frequency spectrum usage between countries and commercial companies, enhance safety by sharing conjunction analyses of satellites, and often advocate for minimizing orbital debris through improved designs and deorbit plans. In general, private space companies seek a stable, predictable space regime—one free of conflict, interference, or damage to the orbital environment. Conflict and occasions when countries pursue military, kinetic, and irreversible actions that deny or impede a commercial service provider's ability to perform its corporate functions are bad for business, potentially resulting in lost revenue, decreased market share, and lower stock price.

In the event of a war extending into space, the commercial space sector is also likely to promote a rapid return to *antebellum* conditions through actions seeking peace and stability. These actions can include calling out "bad" actors—whether state or commercial companies—who act irresponsibly or who conduct unsafe actions, create debris, interfere with satellite communications, degrade space-based services, and the like. Commercial companies, including those participating in sharing space situational awareness information, may increase the transparency of on-orbit activities by publicizing any irresponsible behavior occurring in space. These actions may include providing locating information on the source of communications interference or publicizing the real-time operations of satellites conducting unwelcome or unsafe maneuvering near another satellite. Additionally, space companies can perform digital and cyber forensics on harmful actions impacting space communications and operations, with the purpose of highlighting irresponsible behavior that threatens the free and open use of space by all.

Because of the international nature of large commercial space companies and the fact that many provide services that are transnational, or span multiple countries' borders, commercial companies may have influence with a state's leadership. Many satellite communications companies have licensing agreements or "landing rights" with individual countries to provide communications services within their sovereign territories. Thus, commercial space companies may routinely interact with a country's senior officials and have influence in shaping current and future decisions. In the event of hostile or irresponsible state actions in space, to include jamming of commercial satellite communications, it is likely that a company's Chief Executive Officer or legal counsel will convey their displeasure and desire to resolve the situation on favorable terms.

Ensuring access and use of space

There is an established history of using space for strategic advantage, to include ensuring access to space-enabled capabilities and services. In 1963, the United States launched half a billion whisker-thin copper wires into orbit to effectively create an artificial ionosphere above the Earth.[4] This effort was called Project West Ford, and it was an effort to establish the largest radio antenna in human history, thereby ensuring high-frequency communications in the event a nuclear war. After a failed initial test in 1961, on May 8, 1963 a second launch succeeded in forming a belt of deployed copper dipoles. As the months and years passed, the belt became less effective for its intended purpose, and by early 1966, most of the copper dipoles had reentered the atmosphere as planned.[5] Despite causing a dangerous amount of debris by today's standards, Project West Ford illustrates that there is historical experience of desiring to safeguard space-enabled capabilities and services.

As with the intent of Project West Ford, commercial space providers also help ensure access to and use of space-enabled capabilities and services, which is foundational to space strategy. Incorporating the commercial space sector into an overarching national policy and strategy may, in some cases, allow the governments to forego the lengthy process of designing, acquiring, launching, and operating their own satellites. While it is useful to try to "bin" different commercial activities into either *access* or *use*, there is frequently significant overlap between the two areas. Regardless, considering the two areas separately is indeed useful when developing and implementing a practical strategy that integrates commercial space systems and services.

Commercial space companies will influence the space strategies of states; yet, these companies can have their own corporate strategies as well. A transnational corporation, or any business spanning more than one country, may also have interests in how and what activities are conducted; consequently, these companies may want to develop a space strategy. Space-enabled or space-reliant companies likely do not need to consider explicitly the military instruments of power—including offensive strategy—but they will need to consider a strategy that protects their access to and use of space in times of peace and conflict. Military conflict may potentially impact how, when, and where commercial space operators conduct business or provide services, especially with respect to honoring service level or licensing agreements in potentially non-permissive environments.

Ensuring access

There is a range of commercial space activities related to ensuring access to space. This includes efforts to coordinate frequency spectrum usage, along with measures to develop, manufacture, test, and launch satellites, humans, or other systems into outer space. Activities that facilitate launch vehicle ridesharing or promote hosted payloads on satellites would also be considered as helping ensure access to space.

Additionally, space-enabled or space-reliant corporations will likely want to ensure their access to the most desirable frequency spectrum, thereby providing the most effective and efficient space communications. As a result, these international companies will work with the International Telecommunication Union and national governments to gain approval for using the frequency spectrum and associated landing rights that are optimal for their satellite communication links. If the desired frequency bandwidth is not available, they may lobby to have the frequency assignment changed or else accept less desirable frequency spectrum. Additionally, companies can try to lease the frequency spectrum that is allocated to another country or company, as needed.

U.S. policy notes the need to rely upon domestic commercial space launch as the foundation for the country's access to space.[6] On November 21, 2013, President Obama issued an updated National Space Transportation Policy, which provided guidance to departments and agencies on using commercial and governmental space transportation systems.[7] The goals of the policy are:

- promote and maintain a dynamic, healthy, and efficient domestic space transportation industrial base;
- encourage the U.S. commercial space transportation industry to increase cost effectiveness, foster innovation, and benefit the U.S. economy;
- conduct and promote technology research and development activities to improve the affordability, reliability, performance, safety, and responsiveness of U.S. space transportation capabilities;
- enable the capabilities to support human space transportation activities to and beyond low Earth orbit; and
- foster the development of U.S. commercial spaceflight capabilities serving the emerging non-governmental human spaceflight market.[8]

Per this policy, the National Aeronautics and Space Administration (NASA) is to implement partnerships with the U.S. private sector to develop safe, reliable, and cost effective commercial spaceflight capabilities for the transport of crew and cargo to and from the International Space Station and low Earth orbit.[9] The policy notes that U.S. commercial space transportation capabilities that demonstrate the ability to launch payloads safely and reliably will be allowed to compete for U.S. government missions on a level playing field, consistent with established new entrant certification criteria. Departments and agencies shall use U.S. commercial space transportation capabilities and facilitate multiple U.S. commercial providers of space transportation services across a range of launch vehicle classes, to the maximum extent practicable. Serving as a warning, the policy notes the U.S. government should refrain from conducting space transportation activities that "preclude, discourage, or compete" with U.S. commercial space transportation activities, unless required by national security or public safety.[10]

Because many countries' civil space and national security budgets are expected to see little significant growth in the near-term, some analysts see increased use of public–private partnerships as offering greater efficiency, reducing life-cycle costs, and leading to more innovation.[11] Public–private partnerships are thought to maximize budgets by minimizing any duplication in overhead or mission capabilities. European countries have a long history of taking advantage of the benefits coming from close cooperation between government-sponsored space programs and commercial ones. Similarly, NASA has adapted its acquisition strategies to engage more effectively with commercial firms, especially for the commercial crew and cargo systems being developed by Boeing, Northrop Grumman, Sierra Nevada Corporation, and SpaceX.[12] Some analysts, however, may be skeptical of the commercial market's suitability for space systems and exploration. For instance, some critics argue that long-term science and exploration projects are not an ideal business model for private enterprise because of the requirements of the business cycle, such as quarterly reporting.[13]

Ensuring use

In addition to ensuring access to space, the commercial space sector works to bolster its continued use. Commercial space companies provide capability and services that span most space mission areas and roles. The exception to this would be any specialized military functions where unique requirements preclude the use of commercial systems, because of the sensitivity and critically of the mission performed. Regardless, because the products and services provided by the commercial space sector are wide-ranging with many satellites, associated ground terminals, and data and information networks, military planners should incorporate and integrate commercial capabilities into space strategy.

Commercial space companies have experience operating in a benign and permissive space domain, along with operating in non-permissive environments where attempts are made to intentionally degraded or deny commercial products and services. In the case of satellite communications—which includes television, music, and Internet service—techniques are used to optimize broadcast signals, including beam shaping or multi-beam transmissions to target specific terrestrial areas. These techniques can help mitigate the effects of intentional or unintentional signal interference, and when signal jamming or interference occurs, commercial companies can often determine the location where the interference originated.

Additionally, when it comes to operating through cyber-attacks, this challenge is nothing new for commercial operations. Commercial companies are under the barrage of cyber-attacks on a daily basis, in which state and non-state sponsored hackers attempt to steal intellectual property or degrade satellite or ground station performance. Because commercial companies' livelihood depends on providing its products or services, even when under continuous cyber-attack, many large commercial space companies maintain network and cyber protections as good as, or even better, than most governments.

There are challenges associated with governments using and integrating commercial products and services. These challenges include independent verification and validation of commercial data; tradeoffs in data quality, reliability, availability, and quantity; data sharing policies; and the risk of relying on commercial operators to provide mission critical government data in times of conflict.

Enabling deterrence efforts

Based upon the current commercial space capabilities and emerging technologies, commercial capabilities make it more difficult for malevolent actors to deny access to or use of space. Although not recognized as much as it should among the defense community, the commercial space sector enables more effective deterrence. This includes both deterrence by denial—or dissuasion—and deterrence by punishment (see Chapter 4 for discussions of both).

Commercial space activities aid in improved deterrence by denial. As discussed in Chapter 4, deterrence by denial is achieved through actions causing an adversary to decide that there are credible capabilities to prevent the achieving of potential gains.[14] A deterrence through denial approach seeks to convey the futility of conducting a hostile act, thereby causing a potential adversary's leadership to not pursue a military confrontation in the first place. The commercial space sector contributes to this effort by offering greater space capabilities and services than can be achieve by a single country alone. These commercial capabilities and services include various launch options to get payloads into orbit, substantial number of satellites in constellations performing critical missions, and multiple network paths for space-enabled communications. Commercial launch providers can contribute to improved access to space through responsive launch capabilities, launch rideshares, and the use of hosted payloads on existing or planned satellites. If a commercial remote sensing satellite is damaged during conflict, similar satellites or those of the same constellation can provide the same imagery. Commercial service providers can reroute satellite communications through their own network, or potentially use another company's bandwidth and network if needed. Additionally, commercial capabilities and services may be provided through multi-domain solutions that are diversified and disaggregated (in the parlance of the space professional), and therefore, commercial capabilities and services enable improved space mission assurance and resilience.[15]

Regarding deterrence by punishment efforts, the commercial space sector can play a role, albeit an indirect one. The commercial space sector can do this through improved space situational awareness and space forensics (including digital signatures and multispectral imagery), which may support the attribution process following any hostile or illegal act in space. The commercial space sector already has cyber expertise to support digital forensics efforts and has space situational awareness capability though ground telescopes and other terrestrial tracking systems. If commercial partners support a credible and potentially transparent attribution process, would-be adversaries may decide to act differently if it is known that aggressive and illegal actions will become known. Although the commercial space sector is not to be expected to be involved directly in the use of force following a hostile act, commercial partners may help in providing the information used in determining those responsible. Doing so can help bolster the perceived ability to conduct a legitimate retaliatory response following a hostile attack, which may improve deterrence by punishment efforts.

It should be expected that as the number of commercial satellites grows, the available communications networks increase, and additional multi-domain sources of Earth imaging emerge, the ability to deny services or degrade missions will become more challenging. As a result, deterrence by denial may play a more significant role than deterrence by punishment in the future. This situation can be a good trend, because governments can focus less time and resources on military-related programs that are used in times of conflict and can instead focus on actual commercial services and capabilities that can be used for the benefit of all.

Chinese commercial space

China has recently established a commercial space sector, which is experiencing growth. This move to advance commercial space activities is consistent with the 2016 Chinese Space Activities White Paper, which mentions specifically the actions of "private investors."[16] The policy shift to embrace commercial companies has allowed private investment in the previously closed Chinese space sector. The White Paper addresses a range of mission areas including: launch; meteorological observation; navigation and positioning; telemetry, tracking and command (TT&C); Earth observation; and satellite communications, which all assist "To build China into a space power in all respects...."[17] Under the section heading "System of diverse funding improved," the White Paper says:

> The mechanism for market access and withdrawal has been improved. A list of investment projects in the space industry has been introduced for better management in this regard. Non-governmental capital and other social sectors are encouraged to participate in space-related activities, including scientific research and production, space infrastructure, space information products and services, and use of satellites to increase the level of commercialization of the space industry.[18]

China seeks to pursue commercial space activities because its leadership views, as do those of many other countries, that there is increased demand for the services and capabilities that private space companies can deliver. In noting the country's response to the growing commercial space sector, the White Paper states that China successfully provided a commercial launch service for Turkey's Gokturk-2 Earth observation satellite and is exploring plans to develop a commercial TT&C system.[19]

China has seen notable achievements in its emerging launch market, which presently comprises small launch vehicles. Chinese private launch companies include LinkSpace, OneSpace, iSpace, LandSpace, and ExPace—though ExPace is largely government funded by China Aerospace Science and Industry Corporation and considered only nominally private.[20] In noting the growth in the Chinese private space launch companies, Brian Weeden has commented, "My sense is that these Chinese launch companies are reacting to the same market indicators that all the American launch companies see."[21]

The launch company OneSpace Technologies has enjoyed recent successes. OneSpace, a startup based in Beijing and founded in 2015, became the first Chinese private company to launch successfully its own rocket into space. In May 2018, its 30-foot, single stage solid fuel OS-X rocket launched from a base in northwestern China.[22] It is reported that the vehicle reached an altitude of about 25 miles, flew for about four minutes, reached Mach 5, and covered 170 miles before falling to Earth.[23]

OneSpace is often likened to Elon Musk's SpaceX. The CEO of OneSpace, Shu Chang, says, "Many compare us to SpaceX but to be honest, the gap is more than a little."[24] Although he notes that OneSpace's current situation is very much like where SpaceX was in its early years, the technological differences between the two companies at present are significant. OneSpace's focus, at least initially, is on the low-cost access to space for small payloads market.

Despite being considered private companies, there are some lingering concerns regarding potential undue control or influence by the Chinese government and the People's Liberation Army. Although OneSpace stresses that it is privately owned, the company does have some links to Chinese officials. In U.S. congressional testimony on potential Chinese government-affiliated entities gaining access to U.S. companies through private equity investment, it is stated that OneSpace was founded with direct support from the National Defense Science and Industry Bureau.[25] Also, the company is said to cooperate with Chinese military institutions on research and development and technical services, and OneSpace also has a manufacturing plant in the southwestern city of Chongqing that is partly owned by the local government.[26]

Because of the historical, societal, and cultural differences between Chinese and Western views on the relationship of private companies with government organizations and officials, it should be expected that Chinese commercial space companies will have ties and dealings that look and, at times, behave differently than many other international commercial space companies. Regardless, these expected differences are unlikely to hamper the long-term growth of these Chinese companies because they can tap into the global demand for commercial space products and services.

NewSpace and commercial space investors

In the early 1840s, the third Earl of Rosse designed and built the largest telescope in the world—nicknamed "the Leviathan"—at Birr Castle in central

Ireland. This reflecting telescope remained the largest in the world for over 70 years, and it was used to discover the spiral nature of some of galaxies.[27] While not the only such example, the Leviathan telescope serves as an example how an individual with financial means and an interest in space can play a leadership role in space-related technology and discovery. Today, billionaires such as Elon Musk, Jeff Bezos, and Richard Branson are spending their own fortunes in a similar pursuit of technological advancement and scientific discovery. These billionaires are willing to take great financial risk to see their vision achieved. Many space analysts speculate that these kinds of investors and visionaries have the potential to change the nature of space exploration and improve civil-commercial partnerships in the future.

During the mid-2000s, a new kind of space company emerged.[28] These start-ups took a fresh look at existing space markets, while incorporating the lessons learned from the aerospace industry. These start-up entrepreneurs, some with impressive records of success, developed business plans and sought venture capital from investment firms and angel investors. These so called *NewSpace* companies are presently taking a leading role in space technological development by building the components, materials, and rockets that are deploying a new generation of cell-phone-size satellites into space. In describing *NewSpace*, Joan Johnson-Freese says, "Companies that have become known as NewSpace actors are those largely financed by individuals operating with their own money and so are willing—and able—to take risks."[29] She goes on to observe that NewSpace companies tend to be thinly self-funded or funded by venture capitalists.[30] Moreover, it has been noted that many of these companies, including SpaceX, Blue Origin, and Virgin Orbit, are attracting the best and brightest young minds, which is impacting hiring at traditional aerospace firms.[31] Stability, great benefits, and generous retirement plans are no longer seen as sources of competitive advantage in recruiting young professionals entering the market—inspiration is.

The technology of choice for many of these space start-up companies is the CubeSat, which is ten cubic centimeters, weighs about two pounds, and often costs less than $100,000 to build.[32] Some 60 companies now sell them, allowing governments and companies with relatively little fiscal resources to put a satellite into orbit for agriculture, oil spill, or border monitoring.[33] Some entrepreneurs envision potentially thousands of satellites being part of a mega-constellation in the decade ahead. It has been observed that a launch market has emerged that caters to the CubeSat market.[34]

From analysis of the space start-up market from 2000 to 2016, it is gleaned that hundreds of angel investors, altruists, venture capital firms, private equity firms, corporations, banks, and public markets provided over $16.6 billion to over 140 start-up space companies.[35] Future investment, both in terms of total dollars and number of deals, is expected to increase further year over year. A report by Bryce Space and Technology observes, "The next few years have the potential to transform the start-up space ecosystem."[36] If this transformation is realized in bringing additional capabilities, innovative technologies, and applications to market, it will affect the means available to implement a space strategy. Consequently, New-Space companies and space investors will assist in changing the character of war, while influencing space deterrence, dissuasion, mission assurance, and resilience.

Angel investors

Since 2000, over 140 angel investors have invested in start-up space companies.[37] Typically, angel investors are individuals or families that have accumulated notable wealth and seek high returns by investing in the initial stages of ventures. Investment by angels into start-up space ventures has usually been in the form of straight equity into the company, and investments are commonly from $50,000 to more than $1 million.[38] An angel investor can realize an attractive potential return by investing during a space start-up's nascent stage. Angel investors typically seek to realize their anticipated financial return and exit five to seven years from the date of initial investment. Angels may expect an equity stake in the company as high as 30 to 40 percent in return for their investment.[39] Upon exit, angel investors may expect to receive at least five to ten times their initial investment.[40]

Moreover, there is a high-profile category of angel investor in the commercial space sector, which consists of billionaires and other ultra-high net worth individuals who have a personal stake and interest in NewSpace companies, and the investment level by space billionaires far exceeds those of most angel investors. These billionaires have accrued their wealth through other successful businesses or investments, and they have either founded a space company or invested their own money in a space company. Jeff Bezos, Richard Branson, and Elon Musk are usually the first billionaires mentioned, but they are not the only ones. Of the 1,940 people on Forbes' 2016 World's Billionaires, more than 40 have an affiliation to a space enterprise. This represents about 2 percent of billionaires.[41]

Venture capital

Venture capital firms are groups of investors that invest in start-up, early stage, and growth companies with high revenue potential. In doing so, these firms accept a significant degree of risk. While there is a potential return on investment, there is also high failure rate. In a stark analysis, a research study by a Harvard Business School faculty member finds that three out of four U.S. venture-backed start-ups fail and do not return investors' capital.[42]

Space-oriented venture capital funds are emerging from this investor class. Starburst Accelerator CEO Francois Chopard describes the investment environment of the start-up space ecosystem in saying, "Space technology is today where biotech was about 15 years ago, in terms of potential for startups to flourish."[43] According to the analysis by Bryce Space and Technology, in 2016, Starburst Ventures, an extension of Starburst Accelerator, raised $200 million in funds to invest in 35 start-up space ventures over the following three years. Sera-phim Capital also launched a space technology fund in 2016—the Seraphim Space Fund based in London and focused mostly on U.K.-based companies—which is a $95 million space-focused fund, whose investors include Surrey Sat-ellite Technology, Telespazio, Teledyne, Rolta, First Derivatives, The British Business Bank, the European Space Agency, and the U.K.'s Satellite Applica-tion Catapult.[44] The venture fund represents an interest in future space invest-ments. Additionally, Bessemer Venture Partners announced in 2015 a $1.6 billion fund, BVP IX, to invest in innovative companies and the space sector, to include space companies Terra Bella, Rocket Lab, and Spire.[45]

Private equity

Private equity firms or groups are formed by investors to invest directly in com-panies. Private equity firms typically invest in established companies, instead of start-ups, using substantial transactions that often seek to acquire an entire company or group of related companies. Many types of institutional investors are represented in these firms (e.g., large pension funds), as well as aggregated pools of high net worth individuals. Larger private equity firms are likely to invest between $100 million to $1 billion, usually in the form of equity.[46] They invest sometimes in the form of later stage capital (i.e., later than angel and venture capital investors) or through outright purchase of targeted companies. The larger investment firms, which typically have multi-billion-dollar invest-ment funds, have shown some interest in space over the past 15 years. Firms such as Blackstone, Columbia Capital, Permira, Apax, and Carlyle Group have historically shown an interest for investing in space firms, typically in the tele-communications industry or government contracting.[47]

Emerging capabilities and future trends

Recent entrepreneurial interest and investment in space companies are likely to lead to significant changes in civil, military, and commercial use of and access to space. At times, innovative commercial space services can lead to disputes with governments over regulatory control and licensing requirements. On-orbit satel-lite servicing and asteroid mining are two such areas that may call into question the effectiveness of governmental and international regulations to protect national or global interests. While not an exhaustive list, what follows next are commercial space activities that are emerging or being seriously considered. The list of activities is provided in the context of space strategy and how these com-mercial space activities will affect future political ends that dictate the purpose of conflict, the means available to advance national objectives, and the imple-mentation of practical space strategy.

New launch vehicles

The innovations in the commercial launch sector are improving access to space. The commercial sector is currently developing multiple new launch vehicles across a range of payload classes—small to heavy lift. This development is international in scope and involves established and new launch providers, and this launch market is driven by both governmental and commercial demand to place satellites in orbit. Many commercial launch providers are investing in innovative technologies and processes to lower launch costs and increase competitiveness. Some of the innovative processes include rocket stage and booster reusability, like SpaceX's Falcon Heavy, and the use of lower-cost propellant, like the liquefied natural gas used in Blue Origin's New Glenn vehicles. While technological innovations and new processes are expected to transform the launch market, the market still has persistent challenges including a highly competitive commercial launch market, historically low profit margins, missile technology proliferation concerns, and technology and intellectual property export control risks.

In recent years commercial space companies like Blue Origin and SpaceX have been iteratively working on launch vehicle reusability. It has been observed, however, that only after several years of this activity will it be possible to determine if reusability has led to significant cost savings or not.[48] The next decade will likely see companies and government agencies, working in partnership, continue to develop reusable launch vehicle technologies, with growing operational use and increased launch rates. Additionally, some commercial launch providers are developing new propulsion systems to end current U.S. reliance on Russian RD-180 engines to launch government missions. In 2016, the U.S. Air Force awarded contracts to four companies to develop new first-stage rocket engines, which included Aerojet Rocketdyne for its AR1 kerosene-fueled engine, a team from United Launch Alliance and Blue Origin for the latter's BE-4 methane-fueled engine, SpaceX for its reusable Raptor engine that also uses methane fuel, and Orbital ATK for multiple prototype engines supporting new launch vehicle designs.[49]

There is a hope that large, reusable launch vehicles will lower the cost to access space. This thought has been around for a while, resulting in some asking why the dramatic drop in launch cost has not already been realized.[50] An Air University paper describes the approach under the terminology *Fast Space*, and it is thought that if the U.S. government helps to spur the effort then commercial reusable launch vehicles can transform the launch industry and global access to space. The paper states:

> An infusion of government investment and commitment could jump start a commercial innovation cycle that leads to higher flight rates, decreasing costs, reducing entry barriers for more companies, further increasing demand and higher flight rates, thus reducing costs further. To make Fast Space a reality by breaking the cost equation, the US government will need to jump-start this virtuous cycle.[51]

Yet profit margins for the space launch sector have been historically small, typically about 1 to 3 percent, so time will tell whether innovative launch technologies will, in fact, drive cost down and improve greater access to space.[52]

Many of the advances in the commercial launch market will affect the implementation of space strategy. Commercial launch may benefit reconstitution efforts during times of conflict, in the event mission critical satellites are damaged or destroyed during hostilities. While reconstitution efforts benefit space mission assurance efforts, there is ongoing debate among the national security space community whether it is better to have additional capability on-orbit during peacetime or reconstitute satellites after warfare in space has begun. In the end, a combination of both approaches will likely give the most options for sustained space operations during war.

Use of rideshares and hosted payloads

Rideshares typically use the service of launch consolidators, allowing multiple satellites to deploy from the same launch vehicle. For example, the launch consolidator company Spaceflight Industries seeks to acquire capacity and provide rideshare services on all commercially competitive launch vehicles. These vehicles include the SpaceX Falcon 9, Russian Soyuz, Arianespace Vega, Virgin Orbit LauncherOne, Rocket Lab Electron, Indian PSLV, and others.[53] The company seeks to use a variety of vehicles to launch the spectrum of satellites—CubeSats to large telecommunications satellites—and the company assists in payload integration with the launch vehicle.[54] By combining multiple secondary payloads on launch vehicles servicing a primary payload, launch costs for any rideshares can be a fraction of the cost when compared to having a single, primary payload being launched.

The U.S. government's departments and agencies are currently prohibited from providing rideshares to commercial providers, per the 2013 National Space Transportation Policy. This is because governmental departments and agencies are prohibited from competing with the commercial sector. The policy states that within authorized capacities, departments and agencies shall: "Refrain from conducting United States Government space transportation activities that preclude, discourage, or compete with U.S. commercial space transportation activities, unless required by national security or public safety."[55] Because of this view that the U.S. government should not compete directly with the commercial market, providing rideshare opportunities is solely a commercial activity within the United States.

The hosted payload concept allows the government to fly payloads as additions or attachments to existing, planned commercial satellites, instead of as a free-flying spacecraft requiring dedicated launch services and ground control. The term *hosted payload* has been defined as "the utilization of available power, mass, and space onboard commercial satellites to accommodate additional transponders, instruments or other space-bound items."[56] The hosted portion of the satellite operates independently of the main spacecraft, but shares the satellite's power supply, transponders, and in some cases, ground systems. By offering hosted payload opportunities, it is thought that governmental or other organizations can have reliable and affordable access to space.[57] When the customer is governmental, the company offering hosted payload services can build a trusted relationship with government through a public–private partnership.[58] According the Hosted Payloads Alliance, some of the benefits to the hosted payload customer include:

- shorter time to space because the development of an entire satellite system is not required;
- lower cost because placing a hosted payload on a commercial satellite costs a fraction of the amount of building, launching, and operating a dedicated satellite;
- a more resilient architecture by distributing assets over multiple platforms and locations; and
- increased access to space because the number of commercial launches each year provides multiple opportunities for access to various orbital locations.[59]

Both rideshares and hosted payloads business models support gaining access to space, which improves overall space mission assurance. As commercial launch vehicles increase in size and capacity and satellite hosting capabilities mature, these markets will continue to evolve. There are some persistent challenges for these efforts, however, to include setting launch vehicle and payload interface standards across the industry, and the government's ability to manufacture payloads in a timely fashion to meet fixed commercial launch schedules.

Smaller and more capable satellites

The best is the enemy of the good.

Voltaire[60]

NewSpace start-ups and other space companies are often attempting to exploit current cell phone and microprocessor capability, and consequently, there is a continued desire to place the latest technology into small satellites. Part of these companies' business plans may involve a technology refresh of on-orbit small satellites every 12 to 18 months. Designed and built to incorporate the latest technologies, these small satellites do not necessarily incorporate "space qualified" systems, subsystems, or components as defined by governmental specifications. These satellites do not need to last five or more years on-orbit, since they may be deorbited and replaced well before that time. Weighing less than traditional satellites, small satellites are far cheaper to manufacture. The low-cost satellites using commercial-off-the-shelf technology are said to have led to "democratizing space" for a wide range of participants, from students to entrepreneurs to governments.[61] Hundreds, possibly thousands, of small satellites can potentially be deployed on a single launch vehicle. Moreover, these small satellites do not need to perform the same functions or missions of the most expensive and exquisite satellites. These small satellites that use the latest widely available commercial technology are considered "good enough" for the purposes of the many commercial space start-ups, which conforms to the idea expressed in Voltaire's quote above.

Typically, the size, cost, and capability of satellites depend on their intended function. Some satellites can be held in one's hand, while others, like Hubble Space Telescope, are the size of a school bus. NASA defines small satellites (SmallSats) as spacecraft with a mass less than 180 kilograms and about the size of a large kitchen refrigerator.[62] When considering small satellites, there are sub-categories of SmallSats that are defined by size and mass:

- minisatellite, 100–180 kilograms
- microsatellite, 10–100 kilograms
- nanosatellite, 1–10 kilograms
- picosatellite, 0.01–1 kilograms
- femtosatellite, 0.001–0.01 kilograms.

Furthermore, CubeSats are a class of nanosatellites that use a standardized size and form. The standard CubeSat size uses a "one unit" or "1U" measuring $10 \times 10 \times 10$ centimeters and can be scaled up to larger sizes in modular fashion: 1.5, 2, 3, 6, and even 12U. Originally developed in 1999 by California Poly-technic State University at San Luis Obispo and Stanford University to provide a common platform for education and space exploration, CubeSats have grown into their own industry with government, industry and academia collaborating to increase on-orbit capabilities.[63] CubeSats are viewed to be a cost-effective plat-form for scientific research, technology demonstrations, and advanced mission concepts using large constellations of disaggregated systems.

Rise of the mega-constellations

The increased use of SmallSats has led to the development of large constella-tions of orbiting systems capable of new applications, and it is expected that these constellations will dramatically affect the ability to use space-related prod-ucts and services. Although many of the services are only proposed at present, efforts are underway for these mega-constellations in low Earth orbit to provide global wireless Internet connectivity, cellular phone services, signals collection for global transportation tracking, Earth imaging, and information for data ana-lytics. An advantage of using mega-constellations for many of these missions is the rapid revisit time over the Earth.

The company OneWeb is proposing a constellation of upwards of 900 satel-lites providing global Internet broadband service, and the first customer may be served as early as 2019.[64] OneWeb's satellites will be closer to the Earth allow-ing for better web performance, when compared to satellites providing broad-band from geostationary orbit, like DirectTV.[65] OneWeb understands the technical challenges of operating a constellation, and its company website under-scores the desire to prevent the creation of debris and minimize conjunctions with existing orbital debris. OneWeb satellites are planned to have on-board pro-pulsion to enable maneuvering to prevent conjunctions, and when it comes to end-of-life disposal, the satellites will deorbit automatically, thereby helping to preserve the orbital environment.[66]

SpaceX is planning its own mega-constellation, potentially over 4,000 sat-ellites, each the size of a mini-refrigerator, to provide a global broadband Internet service. The constellation, known as Starlink in U.S. federal filings, will entail thousands of satellites launched on SpaceX launch vehicles being placed in low Earth orbit to provide terrestrial Internet connectivity, thereby potentially bypassing the need for ground-based network infrastructure.[67] The company launched two experimental satellites in February 2018 to help lay the foundation for Starlink.[68] In March 2018, the U.S. Federal Communications Commission (FCC) accepted SpaceX's application to service U.S. customers with its Starlink network, but the FCC conditioned the approval upon an updated deorbit plan, because the large number of satellites envisioned by the constellation exceeds what current U.S. regulatory guidelines consider manageable.[69]

Regulatory and operational challenges of SmallSats and CubeSats that are part of mega-constellations include orbital safety, automated conjunction ana-lysis and collision avoidance, on board propulsion and deorbit capability, debris mitigation, technology refresh rates, dedicated launch vehicles, and radio fre-quency spectrum de-confliction and interference.

Earth imaging and remote sensing

There is a growing market for Earth imagery and associated information coming from satellite-derived data. Because of the growth in computational and techno-logical capabilities, it is expected that there will be a corresponding increase in data and image processing, providing decision-makers with more useful and prompter information. Earth imagery may be multispectral and used to monitor, track, and discern changes in objects, terrain, and weather. This information can make substantial contributions to predictive analytics and forecasting models used in business intelligence, commerce, and trade. Earth imagery can include multispectral imagery from a weather satellite used to forecast or track weather phenomena. This information is useful for the transportation and agricultural industries that are directly impacted by weather changes.

Companies like DigitalGlobe, which trace its origins to the early 1990s, provide Earth imagery that is linked to geospatial locating information. Digital-Globe owns and operates a constellation of high-resolution Earth observation satellites, which are capable of collecting well over one billion square kilometers of quality imagery per year.[70] Because of the increasing demand for Earth imagery, geospatial information, change detection, and data and predictive ana-lytics, other commercial companies are also looking to enter the market.

One of the newer commercial entrants to the sector is Planet, which was founded in 2010 by three former NASA engineers.[71] On February 14, 2017, the company launched 88 Dove satellites into orbit, which was the largest satellite constellation ever to be placed in orbit. This Dove constellation enabled the company to reach its first milestone of imaging all of Earth's landmass every day.[72] At present, the company is operating over 200 Earth observation satellites, which is currently the largest satellite constellation. Planet intends to use the imagery data in the "search engine of the world."[73]

Within the United States, Earth imagery is regulated under the U.S. Commercial Remote Sensing Policy, National Security Presidential Directive-27, which was issued by the George W. Bush administration in April 2003.[74] Among other topics, the White House Fact Sheet on the policy indicates that the U.S. government will:

- rely to the maximum practical extent on U.S. commercial remote sensing space capabilities for filling imagery and geospatial needs for military, intelligence, foreign policy, homeland security, and civil users;
- develop a long-term, sustainable relationship between the government and the U.S. commercial remote sensing space industry;
- provide a timely and responsive regulatory environment for licensing the operations and exports of commercial remote sensing space systems; and
- enable U.S. industry to compete successfully as a provider of remote sensing space capabilities for foreign governments and foreign commercial users, while ensuring appropriate measures are implemented to protect national security and foreign policy.[75]

The policy directs the government to help the U.S. industry compete successfully as a provider of remote sensing space capabilities for both foreign governments and commercial users.[76]

With respect to the last bullet above, the U.S. Commercial Remote Sensing Policy highlights that because of the potential value of its products to an adversary, the operation of a U.S. commercial remote sensing space system requires appropriate measures to address U.S. national security and foreign policy concerns. In such cases, the policy states that U.S. Government may restrict operations of the commercial systems to place conditions on the collection and/or dissemination of certain data and products—e.g., best imagery resolution and most timely delivery to the U.S. government and its approved recipients.[77] On a case-by-case basis, the U.S. government may require additional controls and safeguards for U.S. commercial remote sensing space systems, potentially including them as conditions for U.S government licensing and use of those capabilities. These controls and safeguards shall include, but are not limited to, the unique conditions associated with U.S. government use of commercial remote sensing space systems, including the satellite, ground station, and communications link. Broadly considered, these potential controls are referred to as *shutter control*—or the U.S. government's ability to limit or stop commercial imagery services.[78] There is no reported case, however, of the United States invoking shutter control, but the United States has occasionally bought all commercial imagery at certain times over certain areas such as Afghanistan, also referred to as *checkbook shutter control.*[79]

Since the policy's signing in 2003, much has transpired domestically and internationally. The U.S. commercial sector is not alone in the world market for providing high-resolution imagery. As non-U.S. providers rapidly approach the best that U.S. companies have to offer with respect to resolution and multispectral imagery, U.S. agencies lose their remaining leverage to restrict what imagery is available to the global market place. As James Vedda observes, the U.S. Commercial Remote Sensing Policy of potentially using shutter control will be ineffective in the future if applied only to U.S. commercial remote sensing systems.[80] New commercial satellites, including large constellations of relatively inexpensive satellites, allow for more frequent imaging of terrestrial locations. On-orbit multispectral technology, technology refresh rates of every 12 to 18 months, and on-satellite imagery processing could result in non-U.S. commercial capabilities closing the gap with the most high-end U.S. commercial systems.

For these reasons, U.S. Commercial Remote Sensing Policy will need to evolve as capability and technology advances, if not in word at least in its implementation. Vedda advises that a revised U.S. presidential directive would be beneficial to provide much needed guidance on the U.S. government's treatment of satellite imagery and related hardware, software, and services marketed to commercial and foreign entities.[81] Spatial and spectral resolution, frequency of revisit, timeliness of delivery, and the customer's ability to download directly from a satellite are likely to be key selling points for U.S. commercial remote sensing service providers within the global marketplace. The drafters of any new policy should consider the practices of global competitors before seeking to limit U.S. commercial remote sensing service providers. It is not in the long-term interests of the United States for potential foreign customers look to non-U.S. companies for access to the best imagery, data analytics, and other services.

Data analytics

Based upon Earth observation and geospatial location information, data analytics and other computationally derived products are an increasing market. The benefit of looking at the data analytics sector by itself, versus only at the satellites that perform Earth observation and remote sensing, is that government agencies, commercial industries, and financial markets can focus primarily on the resulting information and meaning ascribed to data and imagery. These groups will not need to develop, build, and launch their own satellites, but governmental or nongovernmental organizations are willing to pay for analysis of the data collected by satellites. The source data used in data analytics can include multispectral imagery or radio frequency signals collection. Through satellites collecting data over a long period—whether years and even decades—information can be gleaned about relative changes on land, at sea, in the air, and in space.

Part of the reason for the growth in data analytics is the ability to store substantial amounts of data (both processed and unprocessed) and new artificial intelligence-like algorithms used for trend and predication analysis based on disparate data sets and sources. Carissa Christensen has commented that the space age has become the data age. Christensen states, "If there is a new space age, this space age is being driven by financial considerations as opposed to technological considerations."[82] Satellite companies are seeking to have imagery and information about everything that is happening on the entire Earth at one time. Christensen observes, "That's extraordinary from a data standpoint. The opportunity to create that data set ... mine it, interpret it and sell the knowledge that comes from it, that's what's different."[83]

The company Planet is making gains in this area. The company plans to use analytical methods and machine learning fed by global, daily Earth imagery to detect and classify objects, identify geographic features, and monitor change over time.[84] The intent is to give governmental, non-governmental, and commercial customers insights and actionable intelligence based upon the most recent imagery available. The company sees its analytical products being used in a number of areas:

- identifying new buildings and roads to update maps and charts, cataloguing urban development, and determining change before and after natural disasters;
- identifying aircraft and monitoring their movement over broad areas to understand economic activity;
- observing maritime vessels to determine "pattern of life" activity, as well as to track specific vessels for law enforcement; and
- detecting deforestation and monitoring land use.[85]

Additive manufacturing

Additive manufacturing techniques are enabling commercial aerospace firms to develop new components for launch systems, satellites, and other payloads at lower development and production costs, lower component weight (resulting in launch cost savings), and on faster production schedules. Additive manufacturing typically uses computer-aided-design software or 3D object scanners to direct hardware to deposit material—layer upon layer—in precise shapes.[86] In contrast, when creating an object by traditional manufacturing means, it is often necessary to remove material through milling, machining, carving, or shaping. Additive manufacturing is expected to impact lower tiers of the supply chain that produce complex parts. Along with the use of commercial off-the-shelf components, additive manufacturing is an important supply chain trend that has the potential to dramatically reduce costs for both the launch and satellite industries. Although the terms *3D printing* and *rapid prototyping* are often used to discuss additive manufacturing, each of those processes is considered a subset of additive manufacturing.

Rapid prototyping techniques have gained popularity among the national security space community, because they are seen as a method of quickly producing hardware for evaluation and testing at the assembly level and below.[87] Designs are drafted and improved upon using engineering software, and then the design results are sent to specialized printers that produce working versions of prototype hardware. Because software is used in the initial design, any needed design changes can be rapidly integrated. This approach is typically much faster and less expensive than the traditional method of delivering specifications and requirements to a machine shop, where prototypes would typically be made.

For example, Rocket Lab, a U.S. aerospace company with operations in New Zealand, has developed and tested the orbital-class Rutherford rocket engine. Notably, the Rutherford engine includes a 3D printed electric turbo-pump. The Rutherford engine was designed to be both high-performing and fast to manufacture. Lachlan Matchett, Vice President of Propulsion at Rocket Lab says, "We can print an entire engine in as little as 24 hours. This allows us to build and launch at unprecedented frequencies to democratize access to space, enabling the creation of crucial orbital infrastructure."[88] Rocket Lab seeks to improve access to space through fast production of affordable, advanced rocket systems and technologies.[89]

The company Made In Space is using a 3D printer aboard the International Space Station (ISS) as a proof of concept and prototyping system. This additive manufacturing device is providing hardware manufacturing services to both NASA and the U.S. National Laboratory onboard.[90] Several other companies are looking to 3D-print components in space to repair other satellites or build structures on-orbit. For instance, NASA announced a project proposal for utilizing public–private partnerships, including Made In Space, Northrop Grumman Corporation, and Oceaneering Space Systems, to develop the necessary technologies and subsystems that will enable the first additive manufacturing, aggregation, and assembly of large and complex systems in space without requiring astronaut extravehicular activity.[91]

While holding great promise for reducing overall cost, weight, and production time, much of additive manufacturing technology in the aerospace sector is still nascent and uncertainties remain. Space is a harsh environment to operate in, especially considering solar radiation, micro-meteorites, and orbital debris. Challenges of additive manufacturing include the dual-use nature of the technologies involved, a viable commercial market to support the effort, and the unknowns regarding quality control and ability to meet "space qualified" standards for exquisite, high-end government systems.

Commercial human spaceflight and space tourism

Multiple companies are developing suborbital and orbital launch and reentry systems capable of flying humans on a commercial basis. Many of these vehicles are for transporting humans to and from the ISS, for point-to-point Earth transportation, or offering microgravity experiences to tourists, researchers, and other customers. Companies like Boeing, Northrop Grumman, Sierra Nevada Corporation, and SpaceX are developing commercial crew and cargo launch vehicles.

Additionally, Blue Origin is pursuing commercial spaceflight for tourism, utilizing its New Shepard vehicle to take passengers into suborbital space inside a crew capsule.[92] Like SpaceX's Falcon series of rockets, New Shepard is designed to launch, land, and be reused. The company hopes to begin flying humans to space in late 2018 or early 2019, starting with its own employees and then followed by paying customers. Blue Origin has yet to announce the price tag for flying passengers, despite reports that a ticket to space could cost between $200,000 and $300,000.[93] Also, Blue Origin has hinted at its lunar ambitions, including the Blue Moon lander capable of placing several tons of cargo on the lunar surface. Jeff Bezos proposed developing Blue Moon as a public–private partnership with NASA. In noting the cost and development advantages of partnership versus each party doing it separately, he remarked, "We can do it a lot faster through a partnership."[94]

Other companies, like Bigelow Aerospace, are developing commercial space habitation pods, allowing for potentially extended space habitation and operations. Bigelow's space station plans are based on soft-bodied modules that launch in a compressed configuration but expand greatly once they reach space. The inflatable structures offer much more habitable volume per unit launch mass and potentially better radiation shielding when compared with traditional aluminum modules. The Bigelow Expandable Activity Module has been attached to the ISS since April 2016. It currently serves as a storage module and is expected to stay berthed to the ISS through at least 2020.[95]

Besides the sheer technological complexity of launching and recovering humans safely, other commercial human spaceflight challenges include: providing recognized safety and governmental regulatory oversight; certification requirements for on-orbit commercial habitats or future space stations; future lunar basing implications with the 1967 Outer Space Treaty; and a tenuous commercial market for an expected expensive service.

Debris removal and mitigation

Orbital debris is a relevant concern for space strategy because debris directly affects the ability to access and use space. Because of growing concerns for space operations safety and preservation of the orbital environment, many companies are looking into options for active debris removal, particularly in low Earth orbit. Of the more than 1,400 functional satellites currently orbiting the Earth, many contend with approximately 500,000 pieces of human-generated space debris in orbit.[96] While the international community has been consistent about advocating for measures to mitigate the generation of debris, the ever-growing number of trackable and un-trackable debris objects remains a concern. Consequently, methods of preventing future debris and removing existing debris are growing areas of investigation.

Surrey Satellite Technology Ltd. has analyzed potential methods for active and passive debris removal.[97] While these approaches appear technologically feasible, time will tell if there is indeed a market to sustain commercial or government-sponsored debris removal activities. Potential solutions to accelerate the deorbit of debris or move it to a different location include:

- a mass driver, like an electromagnetic catapult for debris;
- attaching a satellite tug to debris;
- harpooning debris to relocate it;
- imparting a force via laser to accelerate the deorbit of debris;
- attaching a tether that increases atmospheric drag;
- catching debris with a net;
- using a grappler or robotic arm to remove or relocate debris;
- using a propulsive exhaust plume to expedite smaller debris' deorbit;
- using a "slingsat" to sling-shot debris to expedite deorbit; and
- attaching a deorbit sail to increase atmospheric drag in low Earth orbit.[98]

Astroscale, a private company established in 2013 and headquartered in Singapore, seeks to help with long-term spaceflight safety by developing space debris removal services.[99] Two main initiatives of the company include: orbital debris monitoring using the IDEA OSG-1 25-kilogram microsatellite that will catalogue and characterize small-sized debris in low Earth orbit; and the End-of-Life Service by Astroscale (ELSA) program focusing on spacecraft retrieval and deorbit service for satellite operators.[100]

In the near term, it is expected that governmental agencies will need to take a leadership role in promoting and incentivizing debris removal approaches, including active debris removal by commercial space entities. This leadership is needed because there is, at present, no purely commercial market to sustain debris removal. Governmental agencies have equities in debris removal for operational safety and environmental impact to the orbital regime, and governments can incentivize debris removal by offering tax relief benefits for commercial companies pursuing active and passive debris removal systems. While it is uncertain whether there is a commercial market to sustain commercial debris removal, repurposing larger pieces of debris—although ownership of orbital debris concerns exist—or converting the material into feed stock for 3D printers for on-orbit manufacturing could help offset expenses. Eventually, a stable commercial debris removal market is expected to materialize, and these companies can help pay for their own operations through the recycling or repurposing of orbital debris.

Space situational awareness

In recent years, there has been an increase in commercial space situational awareness (SSA) capability. This growth is significant, because SSA is foundational to all other activities that ensure access to and use of space. Commercial SSA capabilities are enabled by Earth-based systems, including ground-based telescopes and radars.

The company ExoAnalytic Solutions seeks to preserve space as a safe natural resource, through persistent, automated, and real-time SSA solutions.[101] The company defines SSA as the ability to monitor, understand, and predict natural and man-made objects in orbit around the Earth.[102] The company employs a global commercial SSA telescope network with more than 25 observatories and 200 telescopes to track man-made space objects within multiple orbital regimes, including geosynchronous Earth orbits, highly elliptical orbits, and medium Earth orbits.[103]

Using the globally collected SSA information, the company AGI uses software to model, analyze, and visualize objects in space.[104] Its products are used for space catalog maintenance and observation processing, maneuver processing, sensor tasking, conjunction assessment, and web-based visualization.[105] The Space Data Association uses the company's generated SSA information to provide satellite owner/operator information to facilitate safe space operations.

The company LeoLabs uses a worldwide network of ground-based, phased-array radars that provide high-resolution data on objects in low Earth orbit.[106] LeoLabs is a venture-funded company based in Menlo Park, California providing its services to commercial satellite operators, government regulatory and space agencies, and satellite management services firms.[107] Because of the planned rise in SmallSats and CubeSats, the company identified a market for its products and services to mitigate risk of collisions from the increased congestion.[108] The company offers mapping data and services to mitigate the risks of collisions in space, and these services include rapid orbit determination, early operational support, and ongoing orbit awareness.[109]

On-orbit servicing and proximity operations

Commercial firms are pursuing a variety of new business opportunities involving satellite servicing. An important activity enabling satellite servicing is rendezvous and proximity operations (RPO). There have been RPO demonstrations by NASA and the Defense Advanced Research Projects Agency, which are likely to become much more frequent. These operations are relatively complex, with only a few companies or agencies being able to perform them currently.[110] Some RPO activities include docking and undocking maneuvers, which typically would require a cooperative target. If these initiatives are successful, satellite servicing could extend the operational life of many high-end satellites, thereby realizing lifecycle efficiencies and improved return on investment. Multiple companies are planning to offer on-orbit servicing, including seasoned space companies such as Northrop Grumman, MacDonald, Dettwiler, & Associates, and Airbus.[111] Also, space insurance companies, anticipating a new line of underwriting business, have shown interest in these business plans based on proximity operations.

Stuart Eves has proposed several potential commercial areas for on-orbit satellite servicing.[112] These proposed missions include: service life-extension or technology upgrades of operational satellites; relocation of satellites; satellite inclination lowering; satellite right ascension change; on-orbit refueling; robotic maintenance or repair; monitoring deployment of launch partners; testing space situation awareness capabilities; moving malfunctioning, "embarrassing," or defunct payloads into a "graveyard" orbit; disposing of any inconvenient or hazardous orbital debris; orbital slot occupation; security monitoring of existing satellites; and apogee motor failure compensation.[113]

Operational and regulatory challenges of on-orbit servicing and associated proximity operations include: orbital safety; a dynamic licensing and regulatory regime; a viable and sustainable commercial market; common understanding of liability and property rights in space; the dual-use nature of the technologies involved; and the lack of current governmental authority and oversight in this area.

Resource extraction

For decades, commercial companies and international organizations have discussed plans to mine the Moon and asteroids for precious metals and other vital resources. Some of the precious metals mined would be brought back to Earth for use, while other material would be used for fabricating large structures and components in space.

The company Planetary Resources seeks to identify, extract, and refine resources from near-Earth asteroids.[114] Additionally, the company is focusing on finding and extracting sources of water necessary for humans living in space.[115] The company wants to initially identify asteroids that contain the best sources of water, and afterwards, build an asteroid mining facility to harvest water for sustaining human life, and as propellant for spacecraft. The company estimates that there are two trillion tons of frozen water on near-Earth asteroids.

In 2015, the United States passed the U.S. Commercial Space Launch Competitiveness Act, which includes a provision allowing private companies to claim resources in space.[116] The stated purpose of the Act is "To facilitate a pro-growth environment for the developing commercial space industry by encouraging private sector investment and creating more stable and predictable regulatory conditions, and for other purposes."[117] In 2016, Luxemburg established similar legislation that provides private operators rights to resources in space.[118] Under the U.S. Commercial Space Launch Competitiveness Act, American commercial companies can claim material on celestial bodies for commercial purposes. Moreover, the Space Resource Commercial Exploration and Utilization section of the Act states:

> A United States citizen engaged in commercial recovery of an asteroid resource or a space resource under this chapter shall be entitled to any asteroid resource or space resource obtained, including to possess, own, transport, use, and sell the asteroid resource or space resource obtained in accordance with applicable law, including the international obligations of the United States.[119]

Referencing the 2015 Act, Eric Anderson, co-founder of Planetary Resources, remarked, "This is the single greatest recognition of property rights in history."[120]

Of note, while the Outer Space Treaty explains that states cannot make claims of sovereignty on celestial bodies, under the U.S. Commercial Space Launch Competitiveness Act, a U.S. citizen or commercial entity can exploit space resources for financial gain. To address this point of possible contention, the Act explains, "It is the sense of Congress that by the enactment of this Act, the United States does not thereby assert sovereignty or sovereign or exclusive rights or jurisdiction over, or the ownership of, any celestial body."[121] From the perspective of the legislators within the United States, the Act is still compliant with the Outer Space Treaty legal regime.

Challenges for extracting or mining resources in space include the lack of clear government oversight and a common understanding whether the activity is truly compliant with the Outer Space Treaty or not.

Space-based solar power

Space-based solar power is a concept that seeks to use large arrays of solar cells in orbit and then transmit the energy via microwave or laser to a terrestrial receiving station. The potential for space-based solar power has been discussed since the 1960s. Physicist John Mankins' design calls for arranging thin-film mirrors into a bell shape that can redirect sunlight from almost any angle onto a smaller photovoltaic array.[122] This design would, in theory, bring that power from orbit to the terrestrial electrical grid. Peter Garretson has been a long-time advocate of such technology, seeing it as an energy source that is "fully renewable, produces no greenhouse gasses, is not intermittent, has 24-hour availability, could be made-in-America and could scale to all global demand six times over."[123] Blue Origin founder Jeff Bezos has expressed similar sentiment on space-based solar power.

Despite the theoretical potential, a design has yet to emerge that would be obviously affordable and profitable. The launch is one significant cost, including getting a multi-ton system in geosynchronous orbit. Also, energy transmission losses of both microwave and laser energy from orbit to the terrestrial receiving stations are high. In 2009, a startup called Solaren won a much-publicized contract to supply the biggest utility in California with 200 megawatts of power from space starting in 2016.[124] Space Energy, another startup, also generated a lot of excitement for space solar power. Both firms, however, failed to meet their ambitious timelines.[125] The lack of success in this area begs the question of whether the business concept is sound or if other solar cell or alternative energy solutions will be more profitable. That said, if large solar arrays can be manufactured in space using extraterrestrial material for the majority of the mass, that may improve associated launch and manufacturing costs of the required large solar array.

Will the commercial sector be there when needed?

The commercial space sector will play a substantial role in strategy's development. This will be because either the commerce enabled by space-relevant technologies is considered a vital national interest that needs to be protected or because the commercial space sector will provide the means to help achieve a strategy's goals. Presently, the latter seems to be more the case. For many countries, space-based technology, capabilities, and services are interwoven into how their militaries train and fight. Satellite communication, remote sensing, and global positioning services are extensively used during the conduct of normal military operations. While it may be an exaggeration to say some militaries are "dependent" on space-derived services—because militaries often train for the loss of space-enabled capabilities—it is safe to say they have grown more reliant on them.

Because of the dual-use nature of many of the products and services provided by commercial space activities, it will be difficult, at times, to discriminate between purely military and commercial endeavors and associated systems. There may be shared architectures where military-related communications are enabled by commercial satellites. While there are implications that the strategist must consider, the mixing of military and commercial activities is nothing new. Land, sea, and air operations have all had to consider the blending of military and commercial sectors. Space, as well as cyber, will need to consider the means and methods to target and impact negatively commercial activities that may be commingled with military operations to achieve strategic effect.

Will commercial companies be there to support governments in times of war? When looking to fully integrate the commercial space sector into an overarching space strategy, this is a question on the minds of many military service members. The short answer to the question is "Yes." This question and answer are not unique to the space domain. The aerospace, automotive, and shipbuilding industries have a history of providing military products and services during times of conflict. Unless there are conditions beyond their control, commercial companies will seek to honor the terms of service level or licensing agreements, because reneging on contracts would cause the company to lose market share and future revenue. In short, it's bad business not to keep your word. Yet, it must be underscored that commercial companies will support states in times of conflict *if the applicable agreements are in place before the onset of war.*

To ensure that the commercial sector can provide the most benefit in times of conflict, it is necessary that militaries and commercial partners establish trust during peacetime. Only by establishing trusted relationships and sharing information on commercial products, services, and capabilities can a space strategy be implemented effectively and in a practical manner.

Occasionally, government or military personnel may presume that commercial partners and their services will not be available during conflict, believing that the commercial space sector's capabilities cannot operate or withstand a non-permissive or hostile space domain. Such thinking is unfounded. Many commercial space service providers operate in a non-permissive environment every day. Commercial space companies are routinely under cyber-attack, whether by individuals, foreign militaries, or their surrogates. Many commercial space capabilities are more robust and resilient than generally understood by policy-makers and warfighters. Commercial satellite operators have become more resilient because of the various threats—jamming of satellite communications or cyber-attack of networks—they deal with every day. Also, many of the medium-to-large commercial space companies conduct their own research and development to improve upon how they operate under jamming or cyber-attack conditions, and governments can benefit by applying the lessons learned of commercial partners.

To best incorporate innovative commercial space capabilities, companies and governmental organizations should thoroughly understand certain subjects before conflict occurs. These areas of understanding include:

- commercial companies and governmental licensing authorities;
- the implications when commercial assets are employed to support military activities;
- governments considering the ways and means needed to protect commercial space assets when employed to support military operations;
- companies and their shareholders needing to consider the implications of commercial space assets becoming targets for kinetic or non-kinetic attack because of the services provided to governments during hostilities;
- for commercial remote sensing companies—like those falling under U.S. policies and licensing regulations—understanding the potential level of control that licensing nations may exert during hostilities, to include *shutter control*;
- establishing the most effective and efficient communication structure or architecture between governments and commercial partners, to enable the unimpeded flow of data information during peace and conflict; and
- ensuring commercial partners have access to all necessary data and information—whether classified or not—to ensure they are able to provide the agreed-upon products and services during times of war.

Notes

1 Thucydides, *History of the Peloponnesian War* (432 BC), 1.83.2.
2 Department of Defense, *2018 National Defense Strategy of the United States of America: Sharpening the American Military's Competitive Edge* (2018), 3.
3 The White House, *National Security Strategy of the United States of America* (December 2017), 31, www.whitehouse.gov/wp-content/uploads/2017/12/NSS-Final-12–18–2017–0905.pdf
4 William W. Ward and Franklin W. Floyd, "Thirty Years of Space Communications Research and Development at Lincoln Laboratory," in *Beyond the Ionosphere: Fifty Years of Satellite Communication*, ed. Andrew J. Butrica (Washington: National Aeronautics and Space Administration, 1997), 79–81.

5 Ibid.
6 The White House, *National Space Transportation Policy* (November 21, 2013), 1, https://obamawhitehouse.archives.gov/sites/default/files/microsites/ostp/national_ space_transportation_policy_11212013.pdf
7 Ibid.
8 Ibid., 2.
9 Ibid., 3.
10 Ibid., 4.
11 "Government Objectives: Benefits and Risks of PPPs," Public–Private Partnership Legal Resource Center, World Bank Group, last modified October 31, 2016, accessed November 19, 2017, https://ppp.worldbank.org/public-private-partnership/ overview/ppp-objectives
12 Greg Autry, "America's Future in Space is both Commercial and Traditional," *Spacenews*, August 3, 2017, https://spacenews.com/op-ed-americas-future-in-space-is-both-commercial-and-traditional/
13 Ian Ferguson, "Space Exploration Is Best in Hands of NASA, Not Private Sector," *Mic*, October 26, 2015, accessed November 19, 2017, https://mic.com/articles/2267/ space-exploration-is-best-in-hands-of-nasa-not-private-sector#.bYNbTMCBV
14 Paul K. Davis, "Toward Theory for Dissuasion (or Deterrence) by Denial: Using Simple Cognitive Models of the Adversary to Inform Strategy," RAND NSRD WR-1027 (RAND Corporation, January 2014) 2, www.rand.org/content/dam/rand/pubs/ working_papers/WR1000/WR1027/RAND_WR1027.pdf
15 Office of the Assistant Secretary of Defense for Homeland Defense and Global Security, *Space Domain Mission Assurance: A Resilience Taxonomy* (September 2015), http://policy.defense.gov/Portals/11/Space%20Policy/ResilienceTaxonomy WhitePaperFinal.pdf?ver=2016-12-27-131828-623
16 The State Council Information Office of the People's Republic of China, "Status Report From: China National Space Administration" (December 2016), referenced in "White Paper on China's Space Activities in 2016," *SpaceRef*, December 27, 2016, www.spaceref.com/news/viewsr.html?pid=49722
17 Ibid.
18 Ibid.
19 Ibid.
20 Blaine Curcio and Tianyi Lan, "The Rise of China's Private Space Industry," *Spacenews*, May 25, 2018, https://spacenews.com/analysis-the-rise-of-chinas-private-space-industry/
21 Andrew Jones, "Chinese Commercial Launch Sector Nears Takeoff with Suborbital Rocket Test," *Spacenews*, May 15, 2018, https://spacenews.com/chinese-commercial-launch-sector-nears-takeoff-with-suborbital-rocket-test/
22 Curcio and Lan, "The Rise of China's Private Space Industry."
23 Russ Niles, "Chinese Commercial Space Companies Emerge," *AVweb*, May 20, 2018, www.avweb.com/avwebflash/news/Chinese-Commercial-Space-Companies-Emerge-230847-1.html
24 Curcio and Lan, "The Rise of China's Private Space Industry."
25 Jeffrey Z. Johnson, President and CEO, SquirrelWerkz, "Chinese Investment in the United States: Impacts and Issues for Policy Makers," Testimony presented to the U.S.–China Economic and Security Review Commission, 16, www.uscc.gov/sites/ default/files/Johnson_USCC%20Hearing%20Testimony012617_1.pdf
26 Michelle Toh and Serenitie Wang, "OneSpace Launches China's First Private Rocket," *CNNtech*, May 17, 2018, https://money.cnn.com/2018/05/16/technology/ onespace-china-spacex-startup/index.html
27 "The Great Telescope," Birr Castle Gardens and Science Center, accessed August 8, 2018, http://birrcastle.com/telescope-astronomy/

28 "New Kids on the Block: How New Start-Up Space Companies Have Influenced the U.S. Supply Chain" (Bryce Space and Technology, June 2017), https://brycetech.com/downloads/Start_Up_Space_Supply_Chain_2017.pdf
29 Joan Johnson-Freese, *Space Warfare in the 21st Century: Arming the Heavens* (Abingdon: Routledge, 2017), 138.
30 Ibid., 139.
31 Autry, "America's Future in Space is both Commercial and Traditional."
32 Jennifer Alsever, "Space Startups are Booming in the Mojave Desert," *Fortune*, February 20, 2017, http://fortune.com/2017/02/20/space-startups-travel-satellites/
33 Ibid.
34 Ibid.
35 "Start-Up Space: Update on Investment in Commercial Space Ventures 2018" (Bryce Space and Technology, 2018), vi, https://brycetech.com/downloads/Bryce_Start_Up_Space_2018.pdf
36 Ibid.
37 Ibid., 29.
38 Ibid., 6.
39 Ibid., 6.
40 Ibid., 7.
41 "The World's Billionaires," *Forbes Magazine*, December 31, 2016; Nick DeSantis, "Forbes Billionaires List Map: 2016 Billionaire Population by Country," *Forbes*, March 1, 2016, www.forbes.com/sites/nickdesantis/2016/03/01/forbes-billionaires-list-map-2016-billionaire-population-by-country/#771d1643655d
42 Deborah Gage, "The Venture Capital Secret: 3 Out of 4 Start-Ups Fail," *The Wall Street Journal*, September 20, 2012, www.wsj.com/articles/SB10000872396390443 7202045780049804764291 90
43 Lora Kolodny, "Starburst Ventures Closes $200 Million Debut Fund to Back Space Tech Startups," *TechCrunch*, November 29, 2016, https://techcrunch.com/2016/11/29/starburst-ventures-closes-200-million-debut-fund-to-back-space-tech-startups/
44 "Start-Up Space: Update on Investment in Commercial Space Ventures 2018," 8.
45 Ibid., 9.
46 Ibid., 10.
47 Ibid.
48 "New Kids on the Block: How New Start-Up Space Companies Have Influenced the U.S. Supply Chain," 15.
49 Dan Goure, "Why America Needs a New Upper Stage Rocket More than a Russian RD-180 Replacement," *The National Interest*, January 8, 2017, http://nationalinterest.org/blog/the-buzz/why-america-needs-new-upper-stage-rockey-more-russian-rd-180–23982
50 "Fast Space: Leveraging Ultra Low-Cost Space Access for 21st Century Challenges" (Air University, December 22, 2016), 11. www.defensedaily.com/wp-content/uploads/post_attachment/157919.pdf
51 Ibid.
52 "2017 State of the Satellite Industry Report" (Bryce Space and Technology and Satellite Industry Association, June 2017), 9, www.sia.org/wp-content/uploads/2017/07/SIA-SSIR-2017.pdf
53 "Launch Services," Spaceflight, accessed August 8, 2018, http://spaceflight.com/services/launch-services/
54 "Services," Spaceflight, accessed August 8, 2018, http://spaceflight.com/services/
55 The White House, *National Space Transportation Policy*, 4.
56 "Hosted Payloads," SES Government Solutions, accessed August 8, 2018, https://ses-gs.com/solutions/fixed-sat-solutions/hosted-payloads/
57 Ibid.

58 "Benefits of Hosted Payloads," Hosted Payload Alliance, accessed August 8, 2018, www.hostedpayloadalliance.org/Hosted-Payloads/Benefits.aspx

59 Ibid.

60 Voltaire, *La Bégueule, Conte Moral* (Lausanne: Franç Grasset et Comp, 1772).

61 The White House, *National Security Strategy of the United States of America* (December 2017), 31.

62 "What are Smallsats and CubeSats," The National Aeronautics and Space Administration, accessed August 8, 2018, www.nasa.gov/content/what-are-smallsats-and-cubesats

63 Ibid.

64 "OneWeb," OneWeb, accessed August 8, 2018, www.oneweb.world/

65 Ibid.

66 Ibid.

67 Emre Kelly, "SpaceX's Shotwell: Starlink Internet Will Cost about $10 Billion and 'Change the World,'" *Florida Today*, April 26, 2018, www.floridatoday.com/story/tech/science/space/2018/04/26/spacex-shotwell-starlink-internet-constellation-cost-10-billion-and-change-world/554028002

68 Mike Wall, "SpaceX's Prototype Internet Satellites are Up and Running," *Space*, February 22, 2018, www.space.com/39785-spacex-internet-satellites-starlink-constellation.html

69 Caleb Henry, "FCC Approves SpaceX Constellation, Denies Waiver for Easier Deployment Deadline," *Spacenews*, March 29, 2018, https://spacenews.com/us-regulators-approve-spacex-constellation-but-deny-waiver-for-easier-deployment-deadline/

70 "Our Constellation," DigitalGlobe, accessed August 8, 2018, www.digitalglobe.com/about/our-constellation

71 Michael Baylor, "Planet Labs Targets a Search Engine of the World," *NASA Spaceflight*, January 29, 2018, www.nasaspaceflight.com/2018/01/planet-labs-targets-search-engine-world/

72 "Planet Launches Satellite Constellation to Image the Whole Planet Daily," Planet, press release, February 14, 2017, www.planet.com/pulse/planet-launches-satellite-constellation-to-image-the-whole-planet-daily/

73 Baylor, "Planet Labs Targets a Search Engine of the World."

74 The White House, *U.S. Commercial Remote Sensing Policy*, National Security Presidential Directive 27 (April 25, 2003), www.space.commerce.gov/policy/u-s-commercial-remote-sensing-space-policy/

75 Ibid. "Section II. Policy Goal."

76 "U.S. Commercial Remote Sensing Space Policy," Department of Commerce, Office for Space Commerce, accessed August 8, 2018, www.space.commerce.gov/policy/u-s-commercial-remote-sensing-space-policy/

77 James A. Vedda, "Updating National Policy on Commercial Remote Sensing" (The Aerospace Corporation, March 2017), 5, https://aerospace.org/sites/default/files/2018–05/CommercialRemoteSensing_0.pdf

78 Ibid., 8.

79 Peter L. Hays, *Space and Security: A Reference Handbook* (Santa Barbara, CA: ABC-CLIO, LLC, 2001), 39–40.

80 Vedda, "Updating National Policy on Commercial Remote Sensing," 8.

81 Ibid., 7.

82 Quoted in Alan Boyle, "Why Data Analytics is Becoming the Next Frontier for the Commercial Space Industry," *Geekwire*, November 10, 2017, www.geekwire.com/2017/data-analytics-becoming-next-frontier-commercial-space-industry/

83 Ibid.

84 "Making the Move from Imagery to Insights with Planet Analytics," Planet, press release, July 18, 2018, www.planet.com/pulse/planet-analytics-launch/

85 Ibid.
86 "Additive Manufacturing," GE, accessed August 8, 2018, www.ge.com/additive/additive-manufacturing
87 "Rocketlab Reaches 500 Rutherford Engine Test Fires," Rocket Lab, press release, January 31, 2018, www.rocketlabusa.com/news/updates/rocket-lab-reaches-500-rutherford-engine-test-fires/
88 "About Us," Rocket Lab, accessed August 8, 2018, www.rocketlabusa.com/about-us/
89 Ibid.
90 "Additive Manufacturing Facility," Made In Space, accessed August 8, 2018, http://madeinspace.us/projects/amf
91 "Archinaut," Made In Space, accessed August 8, 2018, www.projectarchinaut.com/
92 Elizabeth Howell, "New Shepard: Rocket for Space Tourism," *Space*, April 20, 2018, www.space.com/40372-new-shepard-rocket.html
93 Michael Sheetz, "Blue Origin Will Begin Selling Tickets for Spaceflights after First Crewed Tests, Company Says," *CNBC*, July 12, 2018, www.cnbc.com/2018/07/12/reuters-america-exclusive-jeff-bezos-plans-to-charge-at-least-200000-for-space-rides-sources.html
94 Jeff Foust, "Bezos Outlines Vision of Blue Origin's Lunar Future," *Spacenews*, May 29, 2018, https://spacenews.com/bezos-outlines-vision-of-blue-origins-lunar-future/
95 Mike Wall, "Bigelow Aerospace Launches New Company to Operate Private Space Stations," *Space*, February 20, 2018, www.space.com/39752-bigelow-space-operations-private-space-stations.html
96 Brian Weeden, "Why Outer Space Matters: Brian Weeden on Natural and Human Generated-Threats on Satellites," *Intercross Blog*, October 24, 2016, http://intercrossblog.icrc.org/blog/why-outer-space-matters-brian-weeden-on-natural-and-human-generated-threats-on-satellites
97 Stuart Eves, "On-orbit Servicing, Debris Removal and Emerging Capabilities" (Presentation, Surrey Satellite Technology Limited, London, January 2017).
98 Ibid.
99 "About," Astroscale, accessed August 8, 2018, http://astroscale.com/about
100 "Idea OSG-1," Astroscale, accessed August 8, 2018, http://astroscale.com/services/osg-1; "ELSA-d," Astroscale, accessed August 8, 2018, http://astroscale.com/services/elsa-d
101 "Space Situational Awareness," ExoAnalytics, accessed September 7, 2018, https://exoanalytic.com/space-situational-awareness/
102 Ibid.
103 Ibid.
104 "Products," AGI, accessed September 7, 2018, www.agi.com/products
105 "Enterprise Solutions," AGI, accessed September 7, 2018, www.agi.com/products/enterprise-solutions
106 "About," LeoLabs, accessed September 7, 2018, www.leolabs.space/about
107 Ibid.
108 Ibid.
109 "LeoLabs," LeoLabs, accessed September 7, 2018, www.leolabs.space/
110 Brian Weeden, "Dancing in the Dark Redux: Recent Russian Rendezvous and Proximity Operations in Space," *The Space Review*, October 5, 2015, www.thespacereview.com/article/2839/1
111 Caleb Henry, "Airbus to Challenge SSL, Orbital ATK with New Space Tug Business," *Spacenews*, September 28, 2017, https://spacenews.com/airbus-to-challenge-ssl-orbital-atk-with-new-space-tug-business/; "Orbital ATK on Track to Launch Industry's First Commercial In-Space Satellite Servicing System in 2018," Northrop Grumman, press release, January 24, 2017, https://news.northropgrumman.com/news/features/orbital-atk-on-track-to-launch-industrys-first-commercial-in-space-satellite-servicing-system-in-2018

112 Eves, "On-orbit Servicing, Debris Removal and Emerging Capabilities."
113 Ibid.
114 "Timeline," Planetary Resources, accessed August 8, 2018, www.planetary resources.com/company/timeline/
115 "ARKYD-301," Planetary Resources, accessed August 8, 2018, www.planetary resources.com/missions/arkyd-301/
116 Public Law 114–90, *U.S. Commercial Space Launch Competitiveness Act*, November 25, 2015, www.congress.gov/bill/114th-congress/house-bill/2262/text
117 Ibid.
118 "Luxembourg's New Space Law Guarantees Private Companies the Right to Resources Harvested in Outer Space in Accordance with International Law," the Government of the Grand Duchy of Luxembourg, Ministry of the Economy, press release, November 11, 2016, https://spaceresources.public.lu/content/dam/space resources/press-release/2016/2016_11_11PressReleaseNewSpacelaw.pdf
119 Public Law 114–90, Section 51303.
120 Quoted in "President Obama Signs Bill Recognizing Asteroid Resource Property Rights into Law," Planetary Resources, press release, November 25, 2015, www. planetaryresources.com/2015/11/president-obama-signs-bill-recognizing-asteroid-resource-property-rights-into-law/
121 Public Law 114–90, Section 403.
122 John C. Mankins, "A Fresh Look at Space Solar Power: New Architectures, Concepts and Technologies," *Acta Astronautica* vol. 41 no. 4–10 (1997), 347–359.
123 Peter Garretson, "Better than Paris: Space Solar Power," *The Space Review*, June 19, 2017, www.thespacereview.com/article/3266/1
124 W. Wayt Gibbs, "The Promise of Space-Based Solar Panels," *Discover Magazine*, May 28, 2015, http://discovermagazine.com/2015/july-aug/19-stellar-energy
125 Ibid.

Bibliography

"2017 State of the Satellite Industry Report." Bryce Space and Technology and Satellite Industry Association, June 2017. www.sia.org/wp-content/uploads/2017/07/SIA-SSIR-2017.pdf

"About." Astroscale. Accessed August 8, 2018. http://astroscale.com/about

"About." LeoLabs. Accessed September 7, 2018. www.leolabs.space/about

"About Us." Rocket Lab. Accessed August 8, 2018. www.rocketlabusa.com/about-us/

"Additive Manufacturing." GE. Accessed August 8, 2018. www.ge.com/additive/additive-manufacturing

"Additive Manufacturing Facility." Made In Space. Accessed August 8, 2018. http://madeinspace.us/projects/amf

Alsever, Jennifer. "Space Startups are Booming in the Mojave Desert." *Fortune*. February 20, 2017. http://fortune.com/2017/02/20/space-startups-travel-satellites/

"Archinaut." Made In Space. Accessed August 8, 2018. www.projectarchinaut.com/

"ARKYD-301." Planetary Resources. Accessed August 8, 2018. www.planetary resources.com/missions/arkyd-301/

Autry, Greg. "America's Future in Space is both Commercial and Traditional." *Spacenews*. August 3, 2017. https://spacenews.com/op-ed-americas-future-in-space-is-both-commercial-and-traditional/

Baylor, Michael. "Planet Labs Targets a Search Engine of the World." *NASA Spaceflight*. January 29, 2018. www.nasaspaceflight.com/2018/01/planet-labs-targets-search-engine-world/

"Benefits of Hosted Payloads." Hosted Payload Alliance. Accessed August 8, 2018. www.hostedpayloadalliance.org/Hosted-Payloads/Benefits.aspx

Boyle, Alan. "Why Data Analytics is Becoming the Next Frontier for the Commercial Space Industry." *Geekwire*. November 10, 2017. www.geekwire.com/2017/data-analytics-becoming-next-frontier-commercial-space-industry/

Curcio, Blaine and Tianyi Lan. "The Rise of China's Private Space Industry." *Spacenews*. May 25, 2018. https://spacenews.com/analysis-the-rise-of-chinas-private-space-industry/

Davis, Paul K. "Toward Theory for Dissuasion (or Deterrence) by Denial: Using Simple Cognitive Models of the Adversary to Inform Strategy." RAND NSRD WR-1027. RAND Corporation, January 2014. www.rand.org/content/dam/rand/pubs/working_papers/WR1000/WR1027/RAND_WR1027.pdf

Department of Defense. *2018 National Defense Strategy of the United States of America: Sharpening the American Military's Competitive Edge*. 2018.

DeSantis, Nick. "Forbes Billionaires List Map: 2016 Billionaire Population by Country." *Forbes*. March 1, 2016. www.forbes.com/sites/nickdesantis/2016/03/01/forbes-billionaires-list-map-2016-billionaire-population-by-country/#771d1643655d

"ELSA-d." Astroscale. Accessed August 8, 2018. http://astroscale.com/services/elsa-d

"Enterprise Solutions." AGI. Accessed September 7, 2018. www.agi.com/products/enterprise-solutions

Eves, Stuart. "On-orbit Servicing, Debris Removal and Emerging Capabilities." Presentation. Surrey Satellite Technology Limited. London. January 2017.

"Fast Space: Leveraging Ultra Low-Cost Space Access for 21st Century Challenges." Air University, December 22, 2016. www.defensedaily.com/wp-content/uploads/post_attachment/157919.pdf

Ferguson, Ian. "Space Exploration Is Best in Hands of NASA, Not Private Sector." *Mic*. October 26, 2015. https://mic.com/articles/2267/space-exploration-is-best-in-hands-of-nasa-not-private-sector#.bYNbTMCBV

Foust, Jeff. "Bezos Outlines Vision of Blue Origin's Lunar Future." *Spacenews*. May 29, 2018. https://spacenews.com/bezos-outlines-vision-of-blue-origins-lunar-future/

Gage, Deborah. "The Venture Capital Secret: 3 Out of 4 Start-Ups Fail." *The Wall Street Journal*. September 20, 2012. www.wsj.com/articles/SB10000872396390443720204578004980476429190

Garretson, Peter. "Better than Paris: Space Solar Power." *The Space Review*. June 19, 2017. www.thespacereview.com/article/3266/1

Gibbs, W. Wayt. "The Promise of Space-Based Solar Panels." *Discover Magazine*. May 28, 2015. http://discovermagazine.com/2015/july-aug/19-stellar-energy

Goure, Dan. "Why America Needs a New Upper Stage Rocket More than a Russian RD-180 Replacement." *The National Interest*. January 8, 2017. http://nationalinterest.org/blog/the-buzz/why-america-needs-new-upper-stage-rockey-more-russian-rd-180-23982

"Government Objectives: Benefits and Risks of PPPs." Public–Private Partnership Legal Resource Center, World Bank Group. Last modified October 31, 2016. Accessed November 19, 2017. https://ppp.worldbank.org/public-private-partnership/overview/ppp-objectives

Hays, Peter L. *Space and Security: A Reference Handbook*. Santa Barbara, CA: ABC-CLIO, LLC, 2001.

Henry, Caleb. "FCC Approves SpaceX Constellation, Denies Waiver for Easier Deployment Deadline." *Spacenews*. March 29, 2018. https://spacenews.com/us-regulators-approve-spacex-constellation-but-deny-waiver-for-easier-deployment-deadline/

Henry, Caleb. "Airbus to Challenge SSL, Orbital ATK with New Space Tug Business." *Spacenews*. September 28, 2017. https://spacenews.com/airbus-to-challenge-ssl-orbital-atk-with-new-space-tug-business/

"Hosted Payloads." SES Government Solutions. Accessed August 8, 2018. https://ses-gs.com/solutions/fixed-sat-solutions/hosted-payloads/

Howell, Elizabeth. "New Shepard: Rocket for Space Tourism." *Space*. April 20, 2018. www.space.com/40372-new-shepard-rocket.html

"Idea OSG-1." Astroscale. Accessed August 8, 2018. http://astroscale.com/services/osg-1

Johnson, Jeffrey Z. President and CEO, SquirrelWerkz. "Chinese Investment in the United States: Impacts and Issues for Policy Makers." Testimony presented to the U.S.–China Economic and Security Review Commission. www.uscc.gov/sites/default/files/Johnson_USCC%20Hearing%20Testimony012617_1.pdf

Johnson-Freese, Joan. *Space Warfare in the 21st Century: Arming the Heavens*. Abingdon: Routledge, 2017.

Jones, Andrew. "Chinese Commercial Launch Sector Nears Takeoff with Suborbital Rocket Test." *Spacenews*. May 15, 2018. https://spacenews.com/chinese-commercial-launch-sector-nears-takeoff-with-suborbital-rocket-test/

Kelly, Emre. "SpaceX's Shotwell: Starlink Internet Will Cost about $10 Billion and 'Change the World.'" *Florida Today*. April 26, 2018. www.floridatoday.com/story/tech/science/space/2018/04/26/spacex-shotwell-starlink-internet-constellation-cost-10-billion-and-change-world/554028002

Kolodny, Lora. "Starburst Ventures Closes $200 Million Debut Fund to Back Space Tech Startups." *TechCrunch*. November 29, 2016. https://techcrunch.com/2016/11/29/starburst-ventures-closes-200-million-debut-fund-to-back-space-tech-startups/

"Launch Services." Spaceflight. Accessed August 8, 2018, http://spaceflight.com/services/launch-services/

"LeoLabs." LeoLabs. Accessed September 7, 2018. www.leolabs.space/

"Luxembourg's New Space Law Guarantees Private Companies the Right to Resources Harvested in Outer Space in Accordance with International Law." The Government of the Grand Duchy of Luxembourg, Ministry of the Economy. Press release, November 11, 2016. https://spaceresources.public.lu/content/dam/spaceresources/press-release/2016/2016_11_11PressReleaseNewSpacelaw.pdf

"Making the Move from Imagery to Insights with Planet Analytics." Planet. Press release, July 18, 2018. www.planet.com/pulse/planet-analytics-launch/

Mankins, John C. "A Fresh Look at Space Solar Power: New Architectures, Concepts and Technologies." *Acta Astronautica* vol. 41 no. 4–10 (1997): 347–359.

"New Kids on the Block: How New Start-Up Space Companies Have Influenced the U.S. Supply Chain." Bryce Space and Technology, June 2017. https://brycetech.com/downloads/Start_Up_Space_Supply_Chain_2017.pdf

Niles, Russ. "Chinese Commercial Space Companies Emerge." *AVweb*. May 20, 2018. www.avweb.com/avwebflash/news/Chinese-Commercial-Space-Companies-Emerge-230847-1.html

Office of the Assistant Secretary of Defense for Homeland Defense and Global Security. *Space Domain Mission Assurance: A Resilience Taxonomy*. September 2015. http://policy.defense.gov/Portals/11/Space%20Policy/ResilienceTaxonomyWhitePaperFinal.pdf?ver=2016-12-27-131828-623

"OneWeb." OneWeb. Accessed August 8, 2018. www.oneweb.world/

"Orbital ATK on Track to Launch Industry's First Commercial In-Space Satellite Servicing System in 2018." Northrop Grumman. Press release, January 24, 2017. https://

news.northropgrumman.com/news/features/orbital-atk-on-track-to-launch-industrys-first-commercial-in-space-satellite-servicing-system-in-2018

"Our Constellation." DigitalGlobe. Accessed August 8, 2018. www.digitalglobe.com/about/our-constellation

"Planet Launches Satellite Constellation to Image the Whole Planet Daily." Planet. Press release, February 14, 2017. www.planet.com/pulse/planet-launches-satellite-constellation-to-image-the-whole-planet-daily/

"President Obama Signs Bill Recognizing Asteroid Resource Property Rights into Law." Planetary Resources. Press release, November 25, 2015. www.planetaryresources. com/2015/11/president-obama-signs-bill-recognizing-asteroid-resource-property-rights-into-law/

"Products." AGI. Accessed September 7, 2018. www.agi.com/products

Public Law 114–90. *U.S. Commercial Space Launch Competitiveness Act*. November, 25 2015. www.congress.gov/bill/114th-congress/house-bill/2262/text

"Rocketlab Reaches 500 Rutherford Engine Test Fires." Rocket Lab. Press release, January 31, 2018. www.rocketlabusa.com/news/updates/rocket-lab-reaches-500-rutherford-engine-test-fires/

"Services." Spaceflight. Accessed August 8, 2018. http://spaceflight.com/services/

Sheetz, Michael. "Blue Origin Will Begin Selling Tickets for Spaceflights after First Crewed Tests, Company Says." *CNBC*. July 12, 2018. www.cnbc.com/2018/07/12/reuters-america-exclusive-jeff-bezos-plans-to-charge-at-least-200000-for-space-rides-sources.html

"Space Situational Awareness." ExoAnalytics. Accessed September 7, 2018. https://exo-analytic.com/space-situational-awareness/

"Start-Up Space: Update on Investment in Commercial Space Ventures 2017." Bryce Space and Technology, 2017. https://brycetech.com/downloads/Bryce_Start_Up_Space_2017.pdf

"Start-Up Space: Update on Investment in Commercial Space Ventures 2018." Bryce Space and Technology, 2018. https://brycetech.com/downloads/Bryce_Start_Up_Space_2018.pdf

"The Great Telescope." Birr Castle Gardens and Science Center. Accessed August 8, 2018. http://birrcastle.com/telescope-astronomy/

The State Council Information Office of the People's Republic of China. "Status Report From: China National Space Administration." December 2016.

The White House. *National Security Strategy of the United States of America*. December 2017. www.whitehouse.gov/wp-content/uploads/2017/12/NSS-Final-12-18-2017-0905.pdf

The White House. *National Space Transportation Policy*. November 21, 2013. https://obamawhitehouse.archives.gov/sites/default/files/microsites/ostp/national_space_transportation_policy_11212013.pdf

The White House. *U.S. Commercial Remote Sensing Policy*. National Security Presidential Directive 27. April 25, 2003. www.space.commerce.gov/policy/u-s-commercial-remote-sensing-space-policy/

"The World's Billionaires." *Forbes Magazine*. December 31, 2016.

Thucydides. *History of the Peloponnesian War*. 432 BC.

"Timeline." Planetary Resources. Accessed August 8, 2018. www.planetaryresources.com/company/timeline/

Toh, Michelle and Serenitie Wang. "OneSpace Launches China's First Private Rocket." *CNNtech*. May 17, 2018. https://money.cnn.com/2018/05/16/technology/onespace-china-spacex-startup/index.html

Vedda, James A. "Updating National Policy on Commercial Remote Sensing." The Aerospace Corporation, March 2017. https://aerospace.org/sites/default/files/2018-05/CommercialRemoteSensing_0.pdf

Voltaire. *La Bégueule, Conte Moral.* Lausanne: Franç Grasset et Comp, 1772.

Wall, Mike. "Bigelow Aerospace Launches New Company to Operate Private Space Stations." *Space.* February 20, 2018. www.space.com/39752-bigelow-space-operations-private-space-stations.html

Wall, Mike. "SpaceX's Prototype Internet Satellites are Up and Running." *Space.* February 22, 2018. www.space.com/39785-spacex-internet-satellites-starlink-constellation.html

Ward, William W. and Franklin W. Floyd. "Thirty Years of Space Communications Research and Development at Lincoln Laboratory." *In Beyond the Ionosphere: Fifty Years of Satellite Communication,* edited by Andrew J. Butrica, 79–94. Washington: National Aeronautics and Space Administration, 1997.

Weeden, Brian. "Dancing in the Dark Redux: Recent Russian Rendezvous and Proximity Operations in Space," *The Space Review,* October 5, 2015, www.thespacereview.com/article/2839/1

Weeden, Brian. "Why Outer Space Matters: Brian Weeden on Natural and Human Generated-Threats on Satellites." *Intercross Blog.* October 24, 2016. http://intercross-blog.icrc.org/blog/why-outer-space-matters-brian-weeden-on-natural-and-human-generated-threats-on-satellites

"What are Smallsats and CubeSats." The National Aeronautics and Space Administration. Accessed August 8, 2018. www.nasa.gov/content/what-are-smallsats-and-cubesats

"White Paper on China's Space Activities in 2016." *SpaceRef.* December 27, 2016. www.spaceref.com/news/viewsr.html?pid=49722

9   Looking up and forward

The national security and commercial space sectors are quickly changing. Because of these rapid developments, there is often a desire to predict and forecast what capabilities, services, and actions will occur in the future. When contemplating the desire to predict future events, the sage advice of Arthur C. Clark is useful: "It is impossible to predict the future, and all attempts to do so in any detail appear ludicrous within a few years."[1] The strategist has the problem of prudently and pragmatically considering a future that is unknowable in detail, while needing to reeducate people—and many supposed experts—that their foreseeable futures are nothing of the kind.[2]

The U.S. defense community has historically sought to forecast and predict emerging threats, along with the capabilities needed to combat them. Regarding this desire for certainty about the future, Richard Danzig comments:

> The U.S. military relies on prediction to forecast needs and influence the design of major equipment. A future or futures are envisioned, requirements are deduced and acquisition and design decisions are made and justified accordingly. However, both the experience of the Department of Defense (DOD) and social science literature demonstrate that long-term predictions are consistently mistaken.[3]

Even after investing significant time and resources into prediction and forecasting analysis, errors should be frequently expected.

The strategist needs to acknowledge this reality: predictive failure will occur.[4] Consequently, the strategist's role is to help discern possible, and also implausible, futures to develop fulsome strategies to protect national interests wherever they lie. In providing guidance regarding an unknowable future, Colin Gray observes, "strategists have no choice other than to cope with their unavoidable ignorance as best they may."[5] Policy-makers, strategists, and warfighters should plan across a range of scenarios and potential futures—with the help of timeless wisdom from the strategic theories of Clausewitz, Sun Tzu, Thucydides, and others—to account for the failings of predictive analysis.[6]

Regardless of the prognostication—correct or otherwise—the nature of war is enduring, while its character changes. When seeking to develop strategies and plans for an uncertain future, William Gibson notes: "The future has arrived—it's just not evenly distributed yet."[7] Incorporating Gibson's observation, this chapter addresses current trends and ongoing challenges with how states and non-state organizations think about and operate in space. This discussion will include questions on whether war in space is inevitable, the growing concern with debris and the need for space traffic management, and the necessity to make preparations to cope with great power competition in space. Then, a range of challenges for the United States will be addressed, albeit many of the topics will be germane to the greater spacefaring community. The intent of this chapter's discussion is to provide a useful context for considering the challenges for space strategy during the next three to five years.

Is war in space inevitable?

Occasionally, military and security professionals raise the question of the inevitability of war in space. The context for the question is twofold: the question hints at whether our efforts to prevent future conflict in space is futile; and if war in space is unavoidable, what should be done differently to prepare for this inevitability? To address the first point, J.C. Wylie's general observation is accurate: "*despite whatever effort there may be to prevent it, there may be war.*"[8] The lesson from historical experience is that despite diplomatic endeavors, a sound deterrence strategy, and strategic communications efforts to avoid and prevent conflict, there may indeed be war. The same will hold true of conflict in space.

The concept of the inevitability of conflict extending into space has been discussed for some time. In 1997, the Commander-in-Chief of the U.S. Space Command, General Joseph Ashy, declared that the United States was becoming so dependent on space systems for its armed forces that it had created an enormous incentive for future enemies to exploit this fact. He concludes, "It's politically sensitive, but it's going to happen ... we're going to fight *in* space. We're going to fight *from* space and we're going to fight *into* space...."[9] Similarly in 2002, then U.S. Air Force Colonel and present-day Commander of U.S. Strategic Command, John Hyten, wrote:

> Conflict in space is inevitable. No frontier exploited or occupied by humans has ever been free from strife, but the United States has a chance to mold and shape the resolution of such conflict in the future. Opportunities exist through both formal and informal negotiations to define the commons of space and the rules of the road.[10]

Providing a perspective based upon historical experience and the fundamentals of strategy, Colin Gray observes:

> It is a rule in strategy, one derived empirically from the evidence of two and a half millennia, that anything of great strategic importance to one belligerent, for that reason has to be worth attacking by others. And the greater the

importance, the greater has to be the incentive to damage, disable, capture, or destroy it. In the bluntest of statements: space warfare is a certainty in the future because the use of space in war has become vital.[11]

According to Gray, future warfare will include war in space, at least warfare to contest the control of space.[12]

Today, a question that is often asked within this line of inquiry is whether war between the United States and China is inevitable. This inquiry is relevant to space strategy because any war between the two countries would potentially include war in space. China's rise, especially in space capabilities, in the last few decades is frequently used to emphasize that the days of a bi-polar world during the Cold War and a post-Cold War U.S. unipolar hegemony are a distant memory and that a "new" multipolar world includes "great power competition." Security professionals frequently ask whether China's re-ascendance is a threat to the United States and what should the United States do to address China's economic and military progression?

Specifically, a "Thucydides trap" is discussed within the context of competition between the United States and China, meaning that conflict between the two countries is inevitable. In his compelling description of the Peloponnesian War, Thucydides wrote, "It was the rise of Athens, and the fear that this inspired in Sparta, that made war inevitable."[13] Some have made analogies to Thucydides' statement—although he never used the word *trap* in his work—asking whether the United States fears and mistrusts China in a manner that will lead to war between the two. In 2015, Graham Allison stated that historical evidence shows that the odds of the United States and China going to war were "much more likely than recognized at the moment."[14]

Yet, that is only half the story. As Allison cites, Chinese President Xi Jinping rejected the sentiment of a *trap*, saying "There is no such thing as a so-called Thucydides trap in the world. But should major countries time and again make the mistakes of strategic miscalculation, they might create such traps for themselves."[15] Allison emphasizes that war between the two countries is not inevitable in saying:

> The rise of a 5,000-year-old civilization with 1.3 billion people is not a problem to be fixed. It is a condition—a chronic condition that will have to be managed over a generation. Success will require not just a new slogan, more frequent summits of presidents, and additional meetings of departmental working groups. Managing this relationship without war will demand sustained attention, week by week, at the highest level in both countries. It will entail a depth of mutual understanding not seen since the Henry Kissinger-Zhou Enlai conversations in the 1970s. Most significantly, it will mean more radical changes in attitudes and actions, by leaders and publics alike, than anyone has yet imagined.[16]

Thucydides acknowledged that the human dimension is a causal factor leading to conflict. While people and their assessments contribute to a decision to go to war, war should never be considered unavoidable given the interplay of fear, honor, and interest. A belief in a "Thucydides trap" discounts the rational thinking and free-will that go into weighing the decision to go to war or not.

When considering warfare in space, Xi Jinping's comment above regarding the need to avoid the mistakes of miscalculation rings true. Because of the differing historical, cultural, and societal world-views of China, Russia, and the United States, miscalculation is a definite possibility. As discussed in Chapter 4, these three countries have dissimilar views of deterrence and the efficacy of using military action, and these different views can lead to conflict and military escalation when none was intended. China's view of military action being outside actual war, Russia's perceived use of an escalate-to-deescalate or "unacceptable consequences" approach to deterrence, and the U.S. understanding that military action is justified under the inherent right of self-defense results in different perspectives on deterrence and escalation control. These differences can create a "strategy mismatch," which causes ambiguity and uncertainty, potentially leading to rising tensions and fear among these space powers. By recognizing this mismatch, studying the implications, and promoting dialogue between these countries, many of these differences can be addressed to lessen the possibility of miscalculation and ambiguity. In addressing the different perspectives between China and the United States, Xi Jinping referenced the Chinese saying: "The sun and the moon shine in different ways, yet their brightness is just right for the day and the night respectively."[17] While differences are okay, they must be acknowledged, understood, and communicated to help ensure peace and international stability.

What does all this mean for the space strategist? Strategic history teaches that war in space is likely inevitable—especially in a multipolar environment where space is considered a vital national interest. Despite this potential inevitability, it should never be concluded that conflict in space between two countries is a foregone conclusion. People and the human dimension rule supreme. People may decide to start a war in space, or they may decide to avoid such a conflict. It is a choice.

Growing debris problem and space traffic management

Space debris has been constantly growing since the start of the Space Age. Considered broadly, orbital debris has relevance to the practical implementation of space strategy because debris impacts one's access to and use of space. The United States and Soviet Union contributed to the orbital debris problem during their Space Race, including through anti-satellite (ASAT) testing during the 1980s. Both Russia and the United States are current leaders for creating total orbital debris, and it is estimated that each country has 4,994 and 4,684 uncontrolled objects in space respectively.[18] Of this total debris, rocket bodies—remnants from launches—are a significant concern for orbital safety because of their size and potential for creating additional debris.

During the 2000s there were several noteworthy events creating orbital debris. China's 2007 ground-launched ASAT missile test is estimated to have generated over 3,000 pieces of trackable space debris.[19] While not creating long-term debris, in 2008 the United States used a ship-launched missile defense interceptor—a modified Standard Missile-3—to destroy one of its own defunct intelligence satellites under the name *Operation Burnt Frost*.[20] The purpose for doing so was to prevent the satellite's hydrazine propellant from endangering people upon the satellite's reentry, and the resulting debris caused by the event burned up as it reentered the atmosphere because of the satellite's low altitude.[21] In 2009, an American Iridium commercial communications satellite collided with a defunct Russian military communications satellite, and this conjunction is estimated to have created nearly 2,000 pieces of trackable space debris.[22]

In 1978, National Aeronautics and Space Administration (NASA) scientist Donald Kessler postulated that orbital debris in low Earth orbit can result in a self-sustaining cascading collision, also known as the Kessler Syndrome.[23] This concept first came to NASA's attention in the 1970s when derelict Delta rocket bodies left in orbit "began to explode creating shrapnel clouds."[24] Kessler demonstrated that once the amount of debris in a particular orbit reaches a critical density, collision cascading begins even if no more objects are launched into the orbit. Once this cascading of collisions starts, the risk to satellites and spacecraft grows until eventually it becomes impossible to operate within the orbit.

Presently, orbital debris causes risk of conjunctions for operational satellites and spacecraft. Brian Weeden notes that the more than 1,400 functional satellites currently orbiting the Earth have to contend daily with an estimated 500,000 pieces of human-generated space debris also orbiting the Earth.[25] In 2014, it was reported that satellite operators performed more than 120 maneuvers to change their satellites' paths to reduce the likelihood of colliding with debris.[26] This trend of increasing orbital debris is expected to continue, unless there are incentives or government requirements to clean up or repurpose long-term orbital debris.

The United States has provided guidelines to help minimize future risks of orbital debris. The U.S. Government Orbital Debris Mitigation Standard Practices says:

> *In all operational orbit regimes:* Spacecraft and upper stages should be designed to eliminate or minimize debris released during normal operations. Each instance of planned release of debris larger than 5 mm in any dimension that remains on orbit for more than 25 years should be evaluated and justified on the basis of cost effectiveness and mission requirements.[27]

Many space policy professionals have observed that the 25-year timeframe is too long and impractical in today's commercial space environment where larger constellations with hundreds to thousands of satellites are planned for low Earth orbit. Given the rapid technological advancements in the satellite industry, many commercial companies have shorter operational life cycles for their satellite constellations. Significant technological developments, including those related to Moore's law, mean that satellites become obsolete after only a few years in orbit.

As a result, companies considering the deployment of mega-constellations, like OneWeb, are looking at de-orbiting expired satellites in under five years, not in 25 years as suggested by current U.S. standard practices.[28]

Besides end-of-life satellite disposal, the U.S. Government Orbital Debris Mitigation Standard Practices suggests other safety measures, including: limiting the probability of accidental explosion during and after completion of mission operations; reducing the probability of collision with known objects during a satellite's orbital lifetime; and incorporating satellite survivability against debris collision.[29] Of note, the Inter-Agency Debris Coordination Committee—an international governmental forum for worldwide coordination on man-made and natural debris in space issues—is promoting many of the same considerations contained within the U.S. Government Orbital Debris Mitigation Standard Practices guidelines.[30]

To help address the problem of orbital debris and incentivize commercial companies to repurpose and reuse orbital debris, James Vedda has proposed adding a protocol to the 1975 Convention on Registration of Objects Launched into Outer Space. Article IV of the convention states, "Each State of registry may—from time to time—provide the Secretary-General of the United Nations with additional information concerning a space object carried on its registry." Vedda views this language as permissive to the transfer of ownership of orbital debris or rocket bodies to a third party to enable space salvage or labeling debris as "available for salvage."[31] According to the 1967 Outer Space Treaty's language, states own the launch vehicle, space systems, and any resulting debris from launching or placing a satellite in orbit. Yet Vedda sees modification of the 1975 Registration Convention as a viable method to enable space debris clean-up, without modifying the Outer Space Treaty.

Recently it has been acknowledged that U.S. national and economic interests require an improved domestic space traffic safety governance framework that aims to mitigate and reduce the risk of space incidents due to debris. This need resulted in calls for a civilian agency to oversee Space Traffic Management (STM).[32] Towards this end, in June 2018 President Trump signed the first U.S. National Space Traffic Management Policy.[33] The policy states, "Given the significance of space activities, the United States considers the continued unfettered access to and freedom to operate in space of vital interest to advance the security, economic prosperity, and scientific knowledge of the Nation."[34] The policy document notes that the United States should be a leader in STM and seek to mitigate the effects of debris in space.[35] Also, the policy directs NASA to update the Debris Mitigation Standards, because, in part, of the outdated 25-year rule.[36] With the advancement in space situational awareness (SSA) systems, on-orbit capabilities, and analytical products from commercial partners and governments, the improved knowledge of current satellite and debris locations, along with potential future conjunctions, is thought to help promote safer space operations through STM organizational oversight. As directed by the National Space Traffic Management Policy, the U.S. Department of Commerce is currently moving toward the civil STM role.[37] The policy states:

To facilitate this enhanced data sharing, and in recognition of the need for DoD to focus on maintaining access to and freedom of action in space, a civil agency should, consistent with applicable law, be responsible for the publicly releasable portion of the DoD catalog and for administering an open architecture data repository. The Department of Commerce should be that civil agency.[38]

Making preparations

> The knowledge and skills involved in the preparations will be concerned with the creation, training and maintenance of the fighting forces.
>
> Carl von Clausewitz[39]

Clausewitz wrote on the need to prepare for war, and the advantages of doing so. When considering the activities of warfare, he considered them in two parts:

> To sum up: we clearly see that the activities characteristic of war may be split into two main categories: those that are merely preparations for war, and war proper. The same distinction must be made in theory as well.[40]

The latter area he also referred to as "Theory of the Conduct of War," or in a limited sense, "The Art of War."[41] Yet preparations play into the success that can be achieved during the military operations in conflict; consequently, the preparations that occur ahead of any potential conflict are a critical element of space strategy. It is these preparations, in part, that make the defense the stronger form of war.

For those considering the requisite preparations, it can be a daunting task: preparations must account for an uncertain future and be integrated into the ways and means to achieve political ends. Despite this challenge, the most needed preparations at present for the majority of space powers are:

- better knowledge of what is happening in the space domain;
- improved space mission assurance and resilience; and
- further understanding of the Law of Armed Conflict and Rules of Engagement implications for potential conflict in space.

Space situational awareness

First, fundamental to making adequate preparations is SSA, which U.S. joint doctrine has defined as "cognizance of the requisite current and predictive knowledge of the space environment and the operational environment upon which space operations depend."[42] Effective SSA capabilities will facilitate knowing what on-orbit systems are present, along with their location, capabilities, historical anomalies, operating patterns, and intended use. Such information will facilitate those preparatory measures needed to pit one's strengths against a potential adversary's weakness.

The United States is advancing its SSA capabilities in what is called the Space Fence. This system—based at Kwajalein Atoll in the Marshall Islands—is one part of a Space Surveillance Network. The Space Fence is currently under construction and will become part of an existing radar and optical sensor architecture (terrestrial and space) that will allow better tracking of near-Earth objects in space.[43] System designers believe that the system will improve the Air Force's ability to catalogue space objects from 23,000 to over 200,000 tracked objects.[44] The Space Fence will use ground-based radars to provide uncued detection, tracking, and accurate measurement of space objects, primarily in low Earth orbit.[45] This technology is planned to be operational in 2019, to better enable identification of space objects and improve knowledge of any unexpected satellite maneuvers.[46]

In addition to Space Fence, the United States is improving SSA in geosynchronous orbits through the use of two Geosynchronous Space Situational Awareness Program (GSSAP) satellites.[47] GSSAP satellites operate in the near-geosynchronous orbit regime, supporting U.S. Strategic Command space surveillance operations as a dedicated Space Surveillance Network sensor.[48] The purpose of these satellites, according to U.S. Air Force General John Hyten, is to tell the world that "anything you do in the geosynchronous orbit we will know about. Anything."[49] GSSAP satellites will monitor above and below the geosynchronous belt to capture close-up views of events, to include any hostile actions.[50] These satellites are reported to have enhanced maneuverability to conduct rendezvous and proximity operations for the collection of intelligence.[51] Two GSSAP satellites launched aboard a United Launch Alliance booster from Cape Canaveral on July 2014, and Initial Operational Capability for these satellites was declared in September 2015. Two additional replenishment satellites were launched in August 2016 and accepted into operation in September 2017.[52] Steve Lambakis views GSSAP satellites as providing additional deterrent capability against potential U.S. adversaries because the systems help to bring knowledge of any "bad behavior," while also helping to maintain a safe, secure, and stable space environment.[53]

As illustrated by the terrestrial systems of Space Fence and the on-orbit systems of GSSAP, achieving the requisite level of SSA requires multi-domain solutions. But it also requires a solution benefiting from multi-national and commercial partner participation. Towards this end, more space powers—like Australia—and commercial partners—like ExoAnalytic Solutions—are supporting this effort.[54] Australia's SSA capabilities include an optical tracking observatory in Canberra, and the country's geographical location is beneficial for improving SSA information by tracking satellites from that part of the world. ExoAnalytic Solutions uses a worldwide commercial SSA telescope network that includes over 25 observatories and 200 telescopes tracking objects in various orbital regimes.[55] Because SSA is a global endeavor, information sharing architectures must be designed for including the international community and commercial industry. This means that much of the data and resulting information provided through SSA systems should be releasable and disseminated to many of those participating in the global effort.

Space mission assurance and resilience

Second, improved space mission assurance and resilience measures should be pursued. Clausewitz identified that the defense is the stronger form of war, but only if adequate preparations are taken. These preparations should include incorporating those critical space capabilities required to operate in non-permissive environments or domains of conflict. Consequently, methods of improving mission assurance and resilience should be incorporated broadly across space architectures to operate through and after an attack.

This includes measures meant to promote dissuasion, which may help discourage the initiation of military competition through deterrence by denial.[56] While measures that promote dissuasion help convey the futility of conducting a hostile act—thereby causing a potential adversary's leadership to not pursue a military confrontation in the first place—it incorporates defensive strategies by helping prevent an adversary from taking an objective or achieving its aims.

There are many preparations that enhance mission assurance and resilience measures. Though not an all-inclusive list, some preparations may include: on and off-board protection; hardening; deception; reconstitution and responsive launch capabilities; maneuverability; disaggregation; distribution; and diversification.[57] Less important than using terminology hidden by jargon to describe a function or capability is a recognition that actions should be taken ahead of any conflict that ensure access to and use of space in a hostile and non-permissive environment against a thinking enemy.

With respect to positioning, navigation, and timing (PNT) signals, distribution of these signals enhances mission assurance. This PNT information fuels economic sectors and is used during military operations. Consequently, mission assurance and resilience in this area is important, and space powers are looking at distributing sources of PNT information.[58] This is the case for advancing the use of Multi-Global Navigation Service Signal (GNSS) Receivers.[59] There are multiple PNT constellations: American Global Positioning System (GPS), Russian Global Navigation Satellite System (GLONASS), Chinese Beidou, European Galileo, and Japanese Quasi-Zenith Satellite System (QZSS). Given the concerns of ensuring precision positioning data during wartime, using GNSS receivers is a way to help increase mission assurance and space resilience. Other efforts to provide sources of PNT, some of which are not space-reliant, include dead reckoning inertial navigation systems, fixing information though cellular services and radio beacons, and star trackers.

Law of armed conflict and rules of engagement

Third, a common and practical understanding of what constitutes conflict in space, acceptable behavior, and hostile intent is needed. Space strategy is relatively short on historical experience of warfare when compared to land, sea, and air domains. Consequently, there is little consensus on what constitutes true conflict in space, especially considering the proclivity of reversible and non-kinetic methods of attack. To address this shortcoming, the international community should discuss and debate what constitutes acceptable behavior within context of international treaty and common law. Specifically, more dialogue is needed on what constitutes an "armed attack" (U.N. Charter Article 51 language), "threat or use of force" (U.N. Charter Article 2(4) language), or "hostile act or demonstrated hostile intent" (U.S. Joint Chiefs of Staff Standing Rules of Engagement language).

This international dialogue will help inform a practical understanding of the Law of Armed Conflict (LOAC). As contained within the LOAC, the principle of military necessity calls for using only that degree and kind of force required for the partial or complete submission of the enemy, while considering the minimum expenditure of time, life, and physical resources.[60] Therefore, space strategists and military planners must consider under what conditions the use of military means in space is an appropriate and proportional response in times of conflict. It should be considered ahead of time whether the employment of ground-to-space, space-to-ground, and space-to-space weapons systems are an excessive level of response and whether such an action will be interpreted as escalatory or not. Additionally, this dialogue on the LOAC should include discussions on how and under what conditions anticipatory self-defense—or preemption—in space is considered a legitimate action under international law, considering the principles of necessity, proportionality, and immediacy.[61]

Article 51 of the United Nations Charter states that the use of force as part of the inherent right of self-defense is a legitimate response following an armed attack.[62] To accurately attribute an armed attacked in space to a specific adversary, however, necessitates an SSA capability and architecture that includes a comprehensive and multi-domain space forensics capability. Without significant space forensics capabilities of hostile actions, it seems doubtful an attribution process that is both timely and considered legitimate by the international community is possible. Consequently, future SSA systems need to be able collect and analyze data to gain details following an attack on space systems to support a national-level attribution process.

Making preparations also includes formulating applicable Standing Rules of Engagement relevant to space operations. These rules should include information on when self-defense is warranted for both autonomous and human systems, what is considered space mission essential equipment and infrastructure, and what actions can be taken to defend citizens living, working, or vacationing in space. An international discussion—with the potential of a resulting multinational agreement—on what actions are acceptable in space is a critical first step in developing a common understanding of what constitutes legitimate and illegitimate military conduct in space.

It is noteworthy that the dual-use nature of many space technologies and capabilities makes the application of the LOAC and Rules of Engagement a challenging matter. Technologies that are dual-use can be used for either military or commercial purpose, or even both at the same time. This potential ambiguity makes determining a potential adversary's intent through quantitative measures—such as standoff distances and on-orbit maneuvering—next to impossible. Improvements in SSA may help in this regard, but even substantial improvements may not be enough. Ultimately, determining what is hostile intent or an armed attack in space will likely depend on the broader, geopolitical context. If national security is threatened by actions in space, the LOAC and Rules of Engagement need to inform whether a military response is a legitimate response option or not, even in an operational environment that has ambiguous activities.

Space arms control

Within the national security space community, there is frequent debate regarding the utility of arms control agreements to aid in limiting the spread of weapons deployed into or stationed in space. This debate is lively, in part, because of the different interpretations of the 1967 Outer Space Treaty's (OST) language. Article IV of the treaty affirms:

> States Parties to the Treaty undertake not to place in orbit around the earth any objects carrying nuclear weapons or any other kinds of weapons of mass destruction, install such weapons on celestial bodies, or station such weapons in outer space in any other manner.
>
> The Moon and other celestial bodies shall be used by all States Parties to the Treaty exclusively for peaceful purposes. The establishment of military bases, installations and fortifications, the testing of any type of weapons and the conduct of military manoeuvres on celestial bodies shall be forbidden. The use of military personnel for scientific research or for any other peaceful purposes shall not be prohibited. The use of any equipment or facility necessary for peaceful exploration of the Moon and other celestial bodies shall also not be prohibited.[63]

Except for nuclear weapons and weapons of mass destruction, Article IV does not explicitly prohibit weapons in space or on celestial bodies, yet some view the use of the phrase *peaceful purposes* as conveying that space-based weapons are prohibited. The Arms Control Association notes this view and states, "The treaty repeatedly emphasizes that space is to be used for peaceful purposes, leading some analysts to conclude that the treaty could broadly be interpreted as prohibiting all types of weapons systems, not just WMD, in outer space."[64] To many arms control advocates, agreements are needed to promote international peace and stability in space by curtailing the potential weaponization of space.

Michael Krepon notes the historical successes of arms control agreements within context of future space arms control initiatives. He writes:

> When adversaries perceive common interests to constrain dangerous military technologies, they can focus on preventing tests that are verifiable by national technical means (NTM). Controls on the production of weapon systems incorporating dangerous military technologies are also possible, as

was demonstrated in the Intermediate-range Nuclear Forces Treaty, where production monitoring was accomplished by a combination of on-site inspections and sensors located at and above production facilities. Controls on deployments of military systems incorporating dangerous technologies can also be monitored by cooperative measures and NTM. This is how Washington and Moscow managed to slow down and then downsize their strategic nuclear competition.[65]

Peter Hays observes that arms control measures have been part of great power competition in space since the beginning. This includes the Kennedy administration's space policy focus. The administration took a "two-track" approach to ASAT arms control efforts by deploying a minimum number of ASATs to mitigate the Soviet's orbital nuclear-weapon threat while simultaneously pursing arms-control effort to ban such weapons in space.[66]

Arms control agreements for space prove challenging because of the inspection and verification measures needed as part of such agreements. When conducting inspection and verification for agreements on nuclear weapons, inspectors could be sent to the associated facilities in the host country. For space-based weapons, physical inspection and verification by people will pose challenges. In regard to verifying compliance to an ASAT test ban, using NTM may be possible. Another challenge is that arms control agreements are usually voluntary, begging the question how a space weaponization arms control agreement is to be enforced, although the use of economic sanctions are commonly viewed as a recourse.

In commenting about the utility of arms control for the space domain, at least with ASAT weapons, Steve Lambakis says:

> The danger of declaring or negotiating agreements for peacetime moratoriums on direct-ascent ASATs, for example, is that it would impede the development of capabilities required for space control and limit the development, testing, and potentially the operation of ballistic missile defenses. Moreover, there are very serious definitional and verification problems associated with an ASAT agreement. ASAT weapons can be tested without the target vehicle actually being in orbit. In response to the relative strategic restraint demonstrated by the United States, both Russia and China continue to build up and modernize their ballistic missile and counter-space capabilities.[67]

Hays sets expectations on ASAT arms control measures. He notes that even if all definitional issues and problems with residual and latent ASAT capabilities could be addressed with some sort of controls, it is not clear that such controls would necessarily produce greater stability. It is difficult to divorce offensive and defensive capabilities, and therefore a ground-launched ASAT ban may have the unintended consequences that lead to space-to-Earth weapons system being developed instead, thereby undermining the intended strategic stability.[68] In referencing the various challenges for banning ASAT technology, Hays comments, "Cumulatively, these factors indicate that movement toward effective and stabilizing control of space weapons and missile defenses will remain a daunting challenge and that it is hubris to suggest otherwise."[69]

Krepon recognizes the challenges of arms control for space-related weapons in saying "… deterrence without reassurance is extremely dangerous."[70] He notes that effective deterrence needs accompanying reassurance measures. While acknowledging the challenges of space arms control and associated agreements, Krepon offers one plausible option:

> One place to start, either by tacit or executive agreements with Russia and China, is to stop carrying out hit-to-kill ASAT tests. A kinetic energy ASAT test ban is verifiable and possible because the United States, China, and Russia have already demonstrated this capability, and everyone now recognizes the blowback consequences of explosive debris generation. Agreeing not to carry out such tests would have some symbolic value, as it would demonstrate top-down awareness of the dangers of the current competition. But it would not be reassuring, as it would not constrain competition elsewhere, including ASAT tests designed to miss.[71]

It is expected that arms control measures will remain a challenge for the space domain, even though bans on future on-orbit ASAT testing would be beneficial and potentially verifiable. One enduring problem for arms control is the commingled nature of military and commercial space systems and architectures. This commingling includes the incorporation of many dual-use technologies, hosting government payloads on commercial satellites, and government payloads using rideshares on commercial launch vehicles. This makes it increasingly difficult to conduct inspection and verification to determine compliance with arms control agreements.

Challenges for the United States

This section describes some current challenges facing the Unites States as related to implementing a space strategy. Although the topics are focused on the United States, these challenges are relevant to other space powers as well. These discussion areas include ongoing efforts to reorganize national security space—like a separate Space Force—the desire to maintain a healthy space industrial base, the propensity for mirroring and presentism when formulating strategy, and the rapid growth in SSA capabilities resulting in "turning on the lights" in space.

Space Force

> I am hereby directing the Department of Defense and Pentagon to immediately begin the process necessary to establish a space force as the sixth branch of the armed forces.
>
> President Trump, National Space Council Meeting, June 18, 2018[72]

Within the U.S. national security space community, there have been repeated discussions about whether it is now time to create an independent and separate space service, like the Army, Navy, Marine Corps, and Air Force.[73] Despite the lively and ongoing debate, there seems to be no consensus on the matter. Some may ask, why now? Or is the U.S. Air Force—which has the preponderance of military personnel performing the space mission within the U.S. military—not sufficiently safeguarding or protecting national security interests in space? These questions are complex ones. Peter Hays notes the many historical and ongoing challenges for the U.S. national security space community in saying,

> It is unclear whether the United States will be able to find and follow the best path forward for space strategy, implement the best management and organizational structures for space activities, and sustain the political will needed to continue funding the nearly simultaneous modernizations currently planned.[74]

It is not the intent of the space strategy presented here to address specific organizational or bureaucratic structures, albeit they are indeed important in the execution of space strategy. What will be addressed is if a separate service is advantageous in the long-term and if it is, when change should occur.

When addressing an audience at the Pentagon in August 2018, Vice President Pence reemphasized the administration's agenda in saying, "The time has come to establish the United States Space Force."[75] Pence referenced a Department of Defense report outlining the initial steps to implement the president's guidance to create a Space Force and stated, "This report reviews the national security space activities within the Department of Defense, and it identifies concrete steps that our administration will take to lay the foundation for a new department of the Space Force."[76] The report to Congress addresses four organizational areas where attention and actions are thought to be needed regarding establishing a separate Space Force: Space Development Agency, Space Operations Force, Services and Support, and a new U.S. Space Command.[77]

While the future is unknowable in any detail, empirical historical evidence suggests that future warfare will include conflict in space. Given this, the question is whether a separate space service—a Space Force—is needed. Using a cost versus benefit analysis is one approach to answering the question. Doing so would compare the efficiencies and effectiveness of missions currently performed by the U.S. Air Force and other governmental organizations against potential fiscal requirements and likely space-related threats. A quantitative assessment may prove enlightening, and numerous studies and works have done so in the past.[78] But this approach may miss the human element of the problem. Colin Gray has observed, "Military institutions prepare to fight in the manner that they prefer, unless strategic circumstances or orders from above ... command otherwise."[79] On the importance of service culture in deciding how technology has been employed, Thomas Mahnken observes:

On the other hand, the culture of the U.S. armed services influenced the technologies that they chose to pursue. Technology does not dictate solutions. Rather, it provides a menu of options from which militaries choose. A service's culture, in turn, helps determine which options are more or less attractive.[80]

Therefore, regarding military operations in space, this means a separate space force is likely, in time, to develop its own culture, ethos, and operational style for considering warfare in space. The character of warfare in space is different from the other domains of potential conflict, and a service that focuses on conflict in, from, and through space is more likely to ultimately recognize and adjust to this difference.

In the end, maybe the answer for whether the United States needs a Space Force is the same answer to the question whether the United States needs a Marine Corps. In 1957, Brigadier General Victor Krulak attempted to answer the latter question. He commented,

> The United States does not need a Marine Corps mainly because she has a fine modern Army and a vigorous Air Force.... We [the Marine Corps] exist today—we flourish today—not because of what we know we are, or what we know we can do, but because of what the grassroots of our country believes we are and believes we can do.[81]

So, while the current military services are adequately protecting national security interests in space, maybe the American people now want a Space Force to call their own. Time will have to tell, however.

On the question of timing for a major military reorganization that creates a Space Force, any large organizational changes are best done in peacetime, where time can be more of a luxury. Warfare can be unforgiving in the errors of strategy, operational art, and organizational constructs. Thucydides noted this unforgiving nature, in calling war a "violent teacher."[82] As a result, making large organizational changes—with the potential of failing—and incorporating associated lessons learned is best done absent of open hostilities.

Space industrial base

Since the early establishment of the U.S. commercial sector, there has been a recognition that a healthy domestic industrial base is a national security interest. As a result, there is a common view within the defense community that actions should be taken to ensure a sustainable commercial space sector to develop and produce those systems and capabilities needed to keep the United States a space power frontrunner. There are secondary benefits coming from of a healthy U.S. space industry to include encouraging new industries and creating additional high-technology jobs that lead to greater economic growth.[83] Therefore, a healthy space industrial base benefits the overall U.S. economy as well as being in the nation's interest.[84]

The 2013 U.S. National Space Transportation Policy addresses the desire to promote a healthy and efficient commercial space transportation industrial base, along with government space launch initiatives. The policy directs that departments and agencies should make decisions that consider the health of the U.S. space transportation industrial base, while also pursuing measures such as public–private partnerships and novel acquisition approaches that promote affordability, industry planning, competitive capabilities, infrastructure, and workforce.[85] The National Space Transportation Policy states that maintaining a capability to meet U.S. needs—while also taking the necessary steps to strengthen U.S. competitiveness in the international commercial launch market—is important to ensuring that U.S. space transportation capabilities will be reliable, robust, safe, and affordable in the future.[86]

An important consideration in a healthy space industrial base is the role of the supply chain. The space industry supply chain is the network of companies and suppliers that manufacture and distribute space-related products to customers.[87] Activities within the supply chain include using hardware and materials, components and parts, assemblies, and subsystems to produce a completed system such as a satellite or launch vehicle.[88] Periodically, the U.S. Department of Commerce disseminates surveys within the commercial space sector to uncover potential vulnerabilities in the supply chain, such as sole-source suppliers, reliance on large government programs, and workforce availability.[89] The U.S. defense department is continually assessing industrial base and supply chain risks, to include foreign dependency, sole source, and fragile suppliers, along with suppliers that may be looking to leave the space market.[90]

Because of an ever-expanding globalized commercial space sector, there is increased competition among aerospace companies, both large and small. This competition has led to consolidation in the market to improve production and supply chain efficiencies, which has resulted in many critical space-qualified components having only one supplier—which may reside outside the United States or be manufactured by a non-U.S. parent company. In fact, some critics argue that trying to ensure a healthy U.S. space industrial base or supply chain is a holdover from old ways of thinking. As former Secretary of Defense Robert Gates highlighted, "the U.S. Government has tried to meet post-Cold War challenges and pursue 21st century objectives with processes and organizations designed in the wake of the Second World War."[91] The lesson to be learned is that instead of spending an inordinate amount of time, energy, and money to maintain an indigenous space industrial base, it is more realistic and effective to think in terms of a *global space industrial base*.

Presently, the United States uses and relies on international companies to produce critical space-qualified components and systems. Many of these large, international companies are considered trusted partners because of their proven record of providing critical state-of-the-art components and assemblies for the satellite and space launch industries. Any attempt to ensure all critical technologies are designed and produced solely within the United States will likely result in market inefficiencies and potentially higher prices, without any measurable benefit to national security. Therefore, considering a healthy space industrial base and supply chain should be within a global context.

Mirroring and presentism

For many policy and strategy experts, especially within the United States, there is a common assumption that others think as "we do" and that "tomorrow will be like today only more so."[92] The first part, thinking that others share the same cultural and societal outlooks as Americans, falls under the idea of *mirroring*. Mirroring may manifest itself when deciding upon "most likely" courses of actions during operational planning or determining end states achieved by implementing strategies. During such planning, the underlying assumption is that a potential adversary will think, decide through cost-benefit analysis, or acquiesce to coercive efforts according to one's own mental framework. This thinking is dangerous. Potential adversaries have different cultural, societal, and historical differences that can result in the most basic decisions being starkly different than one's own. This is borne out by previous discussions on the different perspectives of deterrence between China, Russia, and the United States.

The second part, where tomorrow will be like today, falls under the idea of *presentism*. Colin Gray warns of this danger in noting the pervasive use of the phrase "the foreseeable future" among many security and policy experts, because the future is not knowable in any detail.[93] Presentism thinking is also found under other phraseology, like when military leaders are said to be "fighting the last war." For example, just because the last war was an irregular one does not mean the next one will be as well. Andrew Krepinevich states that U.S. defense department bureaucrats would prefer

> no thinking about the future (which implies things might change and they might have to change along with it). To the extent they 'tolerate' such thinking, they attempt to ensure that such thinking results in a world that looks very much like the one for which they have planned.[94]

Gray gives the strategist hope, however. While the strategist must cope with the unavoidable ignorance of foreseeing the future, strategists can also use this understanding to clear away the dead wood of unsound and dangerous assumptions and presumptions.[95]

The U.S. space strategist must make a dedicated and sustained effort to fight a cultural predisposition towards mirroring and presentism. These faulty ways of thinking are biases within the U.S. defense community that will lead to unsound strategy and ineffective operational art. The lesson is others do not necessarily think like oneself, and the future may not look like today. When the space strategist observes this flawed thinking in others, he has the duty to educate and correct the matter.

"Turning on the lights"

As discussed previously with respect to terrestrial and space-based SSA, radar and multispectral technology is advancing rapidly, along with associated change detection, data analytics, and machine learning capabilities. This has resulted in space becoming much more transparent, thereby improving the understanding of a constellation's disposition and knowing a satellite's capabilities. This has led to a figurative "turning on the lights" in space.

This general improvement in SSA and associated non-Earth imaging capabilities has curtailed the ability to maintain previous secrecy in areas once thought to be invisible to public view.[96] In a 2017 survey of members of the U.S. national security space community, James Vedda and Peter Hays posed topics and questions. With respect to non-Earth imaging, respondents commented that security professionals should "Accept that non-Earth imaging is a routine element of future operations;" "The U.S. national security systems have operated with a certain, tacit assumption of privacy. This will no longer be the case in the near future."[97]

The lesson that space professionals should take away is that space is no longer a domain where systems can hide or remain undetected. Instead of worrying about remote sensing, including Earth or non-Earth imaging, the United States should move past this mentality and presume that future activities in space will become more transparent. Therefore, when the operations of space systems are considered sensitive—where their location, disposition, and capability should remain unknown—maintaining privacy will become more challenging to achieve. Space professionals should plan accordingly, where location and movements may become known. If it is desired to keep true capabilities and function unknown, other methods should be pursued, to include hiding in plain sight—or staying visible in a setting that masks presence. Additionally, although many current SSA capabilities are focused on near-Earth orbits, it should be expected that situational awareness of what is happening will eventually push out into cis-lunar space.

Final thoughts

As with any conflict, the decision to willingly pursue war in space should be considered somberly. This is because of the price that war demands. As Gray writes, "War works, it works at a price, and at a price that has to be paid in several currencies: blood, money, influence, honor and reputation."[98] Warfare in space will also demand a price, which may include damage to the natural space environment.

A true conflict in space is likely to have a dramatic effect on the space environment, particularly near-Earth orbits. As the number of spacefaring countries and use of commercial mega-constellations increases, this reality is especially true. Kinetic and non-reversible military actions can cause excessive amounts of orbital debris, which affect negatively access to and use of space within the near-Earth region. Historical experience suggests that these negative effects may last longer than was thought or planned.

The history of using indiscriminate munitions is illustrative of this point. This experience includes using naval contact mines along sea lines of communication, anti-personnel cluster munitions (which many countries have banned) against a terrestrial area target, and improvised explosive devices used in Iraq and Afghanistan along high-traffic areas. These munitions often required extensive time and effort to methodically clear the area affected to regain access and use. A debris-generating war in space is likely to also require significant time, effort, and resources to regain access to and use of the orbital regime. Getting back to an *antebellum* space environment will be difficult or even impossible. Space is a vital interest to the global spacefaring community and many orbital regimes are shared. Consequently, a conflict in space is expected to indiscriminately impact space powers—even those considered neutral.

The development of space strategy is still evolving, with much in the field needing to be better studied and understood. Yet, there are likely more things we know about war in space than things we do not. The enduring nature of war makes this so. It should be expected that war extending or being initiated in space will involve statecraft, strategy, violence, chance, and uncertainty. When considering our nascent understanding of space strategy, it is worth underscoring that what changes over time is far less significant than what remains the same.[99] Robert Kaplan notes this in writing:

> As future crises arrive in steep waves, our leaders will realize that the world is not "modern" or "postmodern" but only a continuation of the ancient—a world that, despite its technologies, the best Chinese, Greek, and Roman philosophers would have understood and known how to navigate.[100]

Because of this continuity in history, the strategist should be optimistic when formulating relevant and practical strategies. The strategist has the entirety of strategic history to draw upon for the development of space strategy. Towards this end, Alfred Thayer Mahan said it simply: "The study of history lies at the foundation of all sound military conclusions and practice."[101] Many space powers have been operating under space strategies for some time, whether acknowledged or not. Everett Dolman hints at this in writing, "The militarization and weaponization of space is not only [a] historical fact, it is an ongoing process."[102] This notion underscores again that the strategist can draw upon this knowledge when refining future strategy, even as the character of space warfare changes. For space powers, the development of space strategy is of critical importance. The stakes are high. The last word on this matter is given to T.E. Lawrence, who offers this charge to the strategist:

> Mankind has had ten-thousand years of experience at fighting and if we must fight, we have no excuse for not fighting well.[103]

Notes

1 Arthur C. Clarke, *Profiles of the Future: An Inquiry into the Limits of the Possible* (New York: Harper and Row, 1962), xi.

2 Colin S. Gray, *Fighting Talk: Forty Maxims on War, Peace, and Strategy* (Westport, CT: Greenwood Publishing, 2007), 156.

3 Richard Danzig, "Driving in the Dark: Ten Propositions about Prediction and National Security" (Center for a New American Security, October 2011), 5, https://s3.amazonaws.com/files.cnas.org/documents/CNAS_Prediction_Danzig.pdf?mtime=20160906081652

4 Ibid., 16.

5 Gray, *Fighting Talk*, 155.

6 Danzig, *Driving in the Dark*, 16.

7 William Gibson quoted in *Cyberpunk* (Documentary), directed by Marianne Trench (Intercon Production, 1990), 12:20.

8 J.C. Wylie, *Military Strategy: A General Theory of Power Control*, with introduction by John B. Hattendorf (New Brunswick, NJ: Rutgers University Press, 1967; reprint, Annapolis, MD: Naval Institute Press, 1989), 66. Emphasis original.

9 Joseph Ashy quoted in Karl Grossman and Judith Long, "Waging War in Space," *The Nation*, December 9, 1999, www.thenation.com/authors/karl-grossman/

10 John E. Hyten, "Sea of Peace or a Theater of War? Dealing with the Inevitable Conflict in Space," *Air and Space Power Journal* vol. 16 no. 3 (Fall 2002), 89, www.dtic.mil/dtic/tr/fulltext/u2/a521811.pdf

11 Colin S. Gray, *Another Bloody Century: Future Warfare* (London: Weidenfeld and Nicolson, 2005), 307.

12 Ibid., 306.

13 Robert B. Strassler, *The Landmark Thucydides: A Comprehensive Guide to the Peloponnesian War* (New York: Free Press, 1996), 16.

14 Graham Allison, "The Thucydides Trap: Are the U.S. and China Headed for War?" *The Atlantic*, September 24, 2015, www.theatlantic.com/international/archive/2015/09/united-states-china-war-thucydides-trap/406756/

15 "Full text of Xi Jinping's speech on China-U.S. relations in Seattle," *Xinhuanet*, September 24, 2015, www.xinhuanet.com/english/2015–09/24/c_134653326.htm

16 Allison, "The Thucydides Trap: Are the US and China Headed for War?"

17 "Full text of Xi Jinping's speech on China–U.S. relations in Seattle."

18 Dave Mosher and Samantha Lee, "More Than 14,000 Hunks of Dangerous Space Junk are Hurtling Around Earth – Here's Who Put It All Up There," *Business Insider*, March 29, 2018, www.businessinsider.com/space-junk-debris-amount-statistics-countries-2018-3

19 Steve Lambakis, "Foreign Space Capabilities: Implications for U.S. National Security" (National Institute for Public Policy, September 2017), 40, www.nipp.org/wp-content/uploads/2017/09/Foreign-Space-Capabilities-pub-2017.pdf

20 Lucas Steinhauser and Scott Thon, "Operation Burnt Frost: The Power of Social Networks," *ASK Magazine* 31 (June 1, 2008), https://appel.nasa.gov/2008/06/01/operation-burnt-frost-the-power-of-social-networks/

21 "Navy Missile hits Dying Spy Satellite, says Pentagon," *CNN*, February 21, 2008, www.cnn.com/2008/TECH/space/02/20/satellite.shootdown/

22 Department of Defense and Office of the Director of National Intelligence, *National Security Space Strategy* (January 2011), 2.

23 Donald J. Kessler and Burton G. Cour-Palais, "Collision Frequency of Artificial Satellites: The Creation of a Debris Belt," *Journal of Geophysical Research* vol. 83 (June 1978), 2637–2646.

24 Ibid., 2645.

25 Brian Weeden, "Why Outer Space Matters: Brian Weeden on Natural and Human Generated-Threats on Satellites," *Intercross Blog*, October 24, 2016, http://intercrossblog.icrc.org/blog/why-outer-space-matters-brian-weeden-on-natural-and-human-generated-threats-on-satellites

26 Ibid.

27 *U.S. Government Orbital Debris Mitigation Standard Practices* (1997), 1, www.iadc-online.org/References/Docu/USG_OD_Standard_Practices.pdf

28 Caleb Henry, "OneWeb Vouches for Higher Reliability of Its Deorbit System," *Spacenews*, July 10, 2017, https://spacenews.com/oneweb-vouches-for-high-reliability-of-its-deorbit-system/

29 *U.S. Government Orbital Debris Mitigation Standard Practices.*

30 "Homepage," Inter-Agency Space Debris Coordination Committee, accessed September 6, 2018, www.iadc-online.org/

31 James Vedda, "Orbital Debris Remediation Through International Engagement," Crowded Space Series Paper #1 (The Aerospace Corporation, March 2017), 6, http://aerospace.wpengine.netdna-cdn.com/wp-content/uploads/2017/09/Debris Remediation.pdf

32 Jeff Foust, "Report Recommends Civil Agency for Space Traffic Management," *Spacenews*, December 28, 2016, http://spacenews.com/report-recommends-civil-agency-for-space-traffic-management/

33 "Space Policy Directive 3 Brings Space Traffic Coordination to Commerce," U.S. Department of Commerce, press release, June 18, 2018, www.commerce.gov/news/press-releases/2018/06/space-policy-directive-3-brings-space-traffic-coordination-commerce

34 The White House, *Space Policy Directive-3, National Space Traffic Management Policy*, Presidential Memoranda (June 18, 2018), www.whitehouse.gov/presidential-actions/space-policy-directive-3-national-space-traffic-management-policy/

35 The White House, *President Donald J. Trump is Achieving a Safe and Secure Future in Space*, Fact Sheets (June 18, 2018), www.whitehouse.gov/briefings-statements/president-donald-j-trump-achieving-safe-secure-future-space/

36 The White House, *Space Policy Directive-3, National Space Traffic Management Policy.*

37 "President Signs Space Traffic Management Policy," Department of Commerce, Office of Space Commerce, press release, June 18, 2018, www.space.commerce.gov/president-signs-space-traffic-management-policy/

38 The White House, *Space Policy Directive-3, National Space Traffic Management Policy.*

39 Carl von Clausewitz, *Vom Kriege*, erster Band (Berlin: Ferdinand Dümmler, 1832), 111.

40 Carl von Clausewitz, *On War*, trans. and eds. Michael Howard and Peter Paret (Princeton, NJ: Princeton University Press, 1989), 131.

41 Ibid., 132.

42 Joint Chiefs of Staff, *Space Operations*, Joint Publication 3–14 (April 10, 2018), GL-6, www.jcs.mil/Portals/36/Documents/Doctrine/pubs/jp3_14.pdf

43 "Space Fence," Lockheed Martin, accessed September 6, 2018, www.lockheedmartin.com/en-us/products/space-fence.html

44 General David J. Buck, Commander, Joint Functional Component Command for Space, Statement to the Committee on Armed Service, House of Representatives, 115th U.S. Congress, March 15, 2016, 5–6.

45 "Space Fence."

46 Debra Werner, "Lockheed Martin Prepares to Turn on U.S. Air Force Space Fence on Kwajalein Atoll," *Spacenews*, May 3, 2018, https://spacenews.com/lockheed-martin-prepares-to-turn-on-u-s-air-force-space-fence-on-kwajalein-atoll/

47 Lambakis, "Foreign Space Capabilities," 45–46.

48 "Geosynchronous Space Situational Awareness Program," U.S. Air Force Space Command Fact Sheet, March 22, 2017, www.afspc.af.mil/About-Us/Fact-Sheets/Article/730802/geosynchronous-space-situational-awareness-program-gssap/

49 Mike Gruss, "Haney: JICSpOC Will Prove U.S. is Prepared for Space Threats," *Spacenews*, August 16, 2016, http://spacenews.com/haney-jicspoc-will-prove-u-s-is-prepared-for-space-threats/

50 James Dean, "Delta IV Blasts Off with Threat-detecting Military Satellites," *Florida Today*, August 19, 2016, www.floridatoday.com/story/tech/science/space/2016/08/19/deltaiv-rocket-blasts-off-air-force-satellites-cape-canaveral-air-force-station-afspc6/88826330/

51 General William Shelton, "The US Future in Space" (The Atlantic Council, Washington, DC, July 23, 2014), www.atlanticcouncil.org/news/transcripts/transcript-the-us-future-in-space

52 "Geosynchronous Space Situational Awareness Program."

53 Lambakis, "Foreign Space Capabilities," 45–46.

54 "Space Surveillance Telescope Australia," M3, accessed September 6, 2018, http://m3eng.com/portfolio/space-surveillance-telescope-australia-2/; "Space Situational Awareness," ExoAnalytic Solutions, accessed September 6, 2018, https://exoanalytic.com/space-situational-awareness/

55 Ian Ritche, "Remote Control Southern Hemisphere SSA Observatory" (EOS Space Systems), www.eos-aus.com/wp-content/uploads/2018/08/Advanced-Maui-Optical-Space-Surveillance-Paper-2013-Southern-Hemisphere-Space-Situational-Awareness-Observatory-1.pdf; "Space Situational Awareness," ExoAnalytic Solutions.

56 Department of Defense, *U.S. Quadrennial Defense Review* (September 30, 2001), 12, http://archive.defense.gov/pubs/qdr2001.pdf

57 Office of the Assistant Secretary of Defense for Homeland Defense and Global Security, *Space Domain Mission Assurance: A Resilience Taxonomy* (September 2015), 3, http://policy.defense.gov/Portals/11/Space%20Policy/ResilienceTaxonomyWhitePaperFinal.pdf?ver=2016-12-27-131828-623

58 Joint Chiefs of Staff, *Space Operations*, Joint Publication 3–14 (April 10, 2018), I–8.

59 G. Manoj Someswar, T.P. Surya Chandra Rao, Dhanunjaya Rao. Chigurukota, "Global Navigation Satellite Systems and their Applications," *International Journal of Software and Web Sciences* 12.136 (2013), www.unoosa.org/documents/pdf/icg/ISWI/IJSWS12–326.pdf

60 U.S. Department of the Navy, *The Commander's Handbook on the Law of Naval Operations*, NWP 1–14M (July 9, 1995), 6–5.

61 Anthony Clark Aren, "International Law and the Preemptive Use of Military Force," *The Washington Quarterly* vol. 26 no. 2 (Spring 2003), 89–103, www.cfr.org/content/publications/attachments/highlight/03spring_arend.pdf

62 United Nations, *Charter of the United Nations and Statue of the International Court of Justice* (San Francisco, June 26, 1945), Chapter 7, Article 51.

63 United Nations General Assembly, resolution 2222 (XXI), *Treaty on Principles Governing the Activities of States in the Exploration and Use of Outer Space, including the Moon and Other Celestial Bodies*, or *The Outer Space Treaty* (1967), Article IV, www.unoosa.org/oosa/en/ourwork/spacelaw/treaties/outerspacetreaty.html

64 Daryl Kimball, "The Outer Space Treaty at a Glance," Fact Sheet (Arms Control Association, August 2017), www.armscontrol.org/factsheets/outerspace

65 Michael Krepon, "Is Space Warfare's Final Frontier?" *Spacenews*, July 24, 2017, https://spacenews.com/op-ed-is-space-warfares-final-frontier/

66 Peter L. Hays, *Space and Security: A Reference Handbook* (Santa Barbara, CA: ABC-CLIO, LLC, 2001), 20.

67 Lambakis, "Foreign Space Capabilities," xiii.

68 Hays, *Space and Security*, 38.
69 Ibid., 77.
70 Krepon, "Is Space Warfare's Final Frontier?"
71 Ibid.
72 The White House, *Remarks by President Trump at a Meeting with the National Space Council and Signing of Space Policy Directive-3*, Remarks (June 18, 2018), www.whitehouse.gov/briefings-statements/remarks-president-trump-meeting-national-space-council-signing-space-policy-directive-3/
73 Jerry Hendrix, "Space: The New Strategic Heartland," *National Review*, June 8, 2018, www.nationalreview.com/2018/06/united-states-needs-space-force-national-security-interest/
74 Hays, *Space and Security*, 52.
75 The White House, *Remarks by Vice President Pence on the Future of the U.S. Military in Space*, Remarks (August 9, 2018), www.whitehouse.gov/briefings-statements/remarks-vice-president-pence-future-u-s-military-space/
76 Ibid.
77 Department of Defense, *Final Report on Organizational and Management Structure for the National Security Space Components of the Department of Defense* (August 9, 2018), 6.
78 For an example of a fulsome study, see *U.S. Commission to Assess United States National Security, Space Management and Organization*, also known as the *Space Commission Report* (January 11, 2001), www.dtic.mil/dtic/tr/fulltext/u2/a404328.pdf
79 Gray, *Fighting Talk*, 103.
80 Thomas G. Mahnken, *Technology and the American War of War Since 1945* (New York: Columbia University Press, 2008), 11.
81 Victor H. Krulak, preface to *First to Fight: An Inside View of the U.S. Marine Corps* (Annapolis, MD: Naval Institute Press, 1984; reprint, Bluejacket Books, 1999).
82 Thucydides, *The Peloponnesian War-The Complete Hobbes Translation*, ed. David Grene (Chicago: The University of Chicago Press, 1989), Book 3, Chapter 82.
83 The White House, *National Space Transportation Policy* (November 21, 2013), 1, https://obamawhitehouse.archives.gov/sites/default/files/microsites/ostp/national_space_transportation_policy_11212013.pdf
84 Ibid., 2.
85 Ibid., 7.
86 Ibid., 1.
87 "New Kids on the Block: How New Start-Up Space Companies Have Influenced the U.S. Supply Chain" (Bryce Space and Technology, June 2017), 6, https://brycetech.com/downloads/Start_Up_Space_Supply_Chain_2017.pdf
88 Ibid.
89 Ibid.
90 Aaron Mehta, "Industrial Base War-gaming: Pentagon Wants Companies to Find Supply Chain Weaknesses," *DefenseNews*, September 28, 2017, www.defensenews.com/smr/equipping-the-warfighter/2017/09/28/industrial-base-wargaming-pentagon-wants-companies-to-find-supply-chain-weaknesses/
91 Robert M. Gates, Secretary of Defense, Opening statement to the Armed Services Committee, House of Representatives, April 15, 2008, http://archive.defense.gov/Speeches/Speech.aspx?SpeechID=1272
92 Gray, *Fighting Talk*, 156.
93 Ibid., 155.
94 Andrew Krepenivich quoted in Danzig, "Driving in the Dark," 12.
95 Gray, *Fighting Talk*, 156.
96 James A. Vedda and Peter L. Hays, "Major Policy Issues in Evolving Global Space Operations" (The Mitchell Institute for Aerospace Studies, February 2018), 24,

www.aerospace.org/publications/policy-papers/major-policy-issues-in-evolving-global-space-operations/
97   Ibid., 47.
98   Gray, *Fighting Talk*, 17.
99   Ibid., 149.
100 Robert D. Kaplan, *Warrior Politics: Why Leadership Demands a Pagan Ethos* (New York: Random House, 2002), vii.
101 Alfred Thayer Mahan, *Armaments and Arbitration or the Place of Force in the International Relations of States* (New York: Harper and Brothers Publishers, 1912), 206.
102 Everett C. Dolman, *Astropolitik: Classical Geopolitics in the Space Age* (London: Frank Cass, 2002), 5.
103 T.E. Lawrence, as attributed in John Hunt, *The Boy Who Could Keep a Swan in His Head* (Cape Town: Penguin Random House, 2018), Chapter 1.

Bibliography

Allison, Graham. "The Thucydides Trap: Are the U.S. and China Headed for War?" *The Atlantic*. September 24, 2015. www.theatlantic.com/international/archive/2015/09/united-states-china-war-thucydides-trap/406756/

Aren, Anthony Clark. "International Law and the Preemptive Use of Military Force." *The Washington Quarterly* vol. 26 no. 2 (Spring 2003): 89–103. www.cfr.org/content/publications/attachments/highlight/03spring_arend.pdf

Buck, David J. Commander, Joint Functional Component Command for Space. Statement to the Committee on Armed Service, House of Representatives, 115th U.S. Congress. March 15, 2016.

Clarke, Arthur C. *Profiles of the Future: An Inquiry into the Limits of the Possible*. New York: Harper and Row, 1962.

Clausewitz, Carl von. *On War*. Translated and edited by Michael Howard and Peter Paret. Princeton, NJ: Princeton University Press, 1989.

Clausewitz, Carl von. *Vom Kriege*, erster Band. Berlin: Ferdinand Dümmler, 1832.

Danzig, Richard. "Driving in the Dark: Ten Propositions about Prediction and National Security." Center for a New American Security, October 2011. https://s3.amazonaws.com/files.cnas.org/documents/CNAS_Prediction_Danzig.pdf?mtime=20160906081652

Dean, James. "Delta IV Blasts Off with Threat-detecting Military Satellites." *Florida Today*. August 19, 2016. www.floridatoday.com/story/tech/science/space/2016/08/19/deltaiv-rocket-blasts-off-air-force-satellites-cape-canaveral-air-force-station-afspc6/88826330/

Department of Defense. *Final Report on Organizational and Management Structure for the National Security Space Components of the Department of Defense*. August 9, 2018.

Department of Defense. *U.S. Quadrennial Defense Review*. September 30, 2001. http://archive.defense.gov/pubs/qdr2001.pdf

Department of Defense and Office of the Director of National Intelligence. *National Security Space Strategy*. January 2011.

Dolman, Everett C. *Astropolitik: Classical Geopolitics in the Space Age*. London: Frank Cass, 2002.

Foust, Jeff. "Report Recommends Civil Agency for Space Traffic Management." *Spacenews*. December 28, 2016. http://spacenews.com/report-recommends-civil-agency-for-space-traffic-management/

"Full text of Xi Jinping's speech on China-U.S. relations in Seattle." *Xinhuanet*. September 24, 2015. www.xinhuanet.com/english/2015-09/24/c_134653326.htm

Gates, Robert M. Secretary of Defense. Opening Statement to the Armed Services Committee, House of Representatives. April 15, 2008. http://archive.defense.gov/Speeches/Speech.aspx?SpeechID=1272

"Geosynchronous Space Situational Awareness Program." U.S. Air Force Space Command. Fact Sheet. March 22, 2017. www.afspc.af.mil/About-Us/Fact-Sheets/Article/730802/geosynchronous-space-situational-awareness-program-gssap/

Gray, Colin S. *Fighting Talk: Forty Maxims on War, Peace, and Strategy*. Westport, CT: Greenwood Publishing, 2007.

Gray, Colin S. *Another Bloody Century: Future Warfare*. London: Weidenfeld and Nicolson, 2005.

Grossman, Karl and Judith Long. "Waging War in Space." *The Nation*. December 9, 1999. www.thenation.com/authors/karl-grossman/

Gruss, Mike. "Haney: JICSpOC Will Prove U.S. is Prepared for Space Threats." *Spacenews*. August 16, 2016. http://spacenews.com/haney-jicspoc-will-prove-u-s-is-prepared-for-space-threats/

Hays, Peter L. *Space and Security: A Reference Handbook*. Santa Barbara, CA: ABC-CLIO, LLC, 2001.

Hendrix, Jerry. "Space: The New Strategic Heartland." *National Review*. June 8, 2018. www.nationalreview.com/2018/06/united-states-needs-space-force-national-security-interest/

Henry, Caleb. "OneWeb Vouches for Higher Reliability of Its Deorbit System." *Spacenews*. July 10, 2017. https://spacenews.com/oneweb-vouches-for-high-reliability-of-its-deorbit-system/

"Homepage." Inter-Agency Space Debris Coordination Committee. Accessed September 6, 2018. www.iadc-online.org/

Hunt, John. *The Boy Who Could Keep a Swan in His* Head. Cape Town: Penguin Random House, 2018.

Hyten, John E. "Sea of Peace or a Theater of War? Dealing with the Inevitable Conflict in Space." *Air and Space Power Journal* vol. 16 no. 3 (Fall 2002): 78–92. www.dtic.mil/dtic/tr/fulltext/u2/a521811.pdf

Joint Chiefs of Staff. *Space Operations*. Joint Publication 3–14. April 10, 2018. www.jcs.mil/Portals/36/Documents/Doctrine/pubs/jp3_14.pdf

Kaplan, Robert D. *Warrior Politics: Why Leadership Demands a Pagan Ethos*. New York: Random House, 2002.

Kessler, Donald J. and Burton G. Cour-Palais. "Collision Frequency of Artificial Satellites: The Creation of a Debris Belt." *Journal of Geophysical Research* vol. 83 (June 1978): 2637–2646.

Kimball, Daryl. "The Outer Space Treaty at a Glance." Fact Sheet. Arms Control Association, August 2017. www.armscontrol.org/factsheets/outerspace

Krepon, Michael. "Is Space Warfare's Final Frontier?" *Spacenews*. July 24, 2017. https://spacenews.com/op-ed-is-space-warfares-final-frontier/

Krulak, Victor H. Preface to *First to Fight: An Inside View of the U.S. Marine Corps*. Annapolis, MD: Naval Institute Press, 1984; reprint, Bluejacket Books, 1999.

Lambakis, Steve. "Foreign Space Capabilities: Implications for U.S. National Security." National Institute for Public Policy, September 2017. www.nipp.org/wp-content/uploads/2017/09/Foreign-Space-Capabilities-pub-2017.pdf

Mahan, Alfred Thayer. *Armaments and Arbitration or the Place of Force in the International Relations of States*. New York: Harper and Brothers Publishers, 1912.

Mahnken, Thomas G. *Technology and the American War of War Since 1945*. New York: Columbia University Press, 2008.

Mehta, Aaron. "Industrial Base War-gaming: Pentagon Wants Companies to Find Supply Chain Weaknesses." *DefenseNews*. September 28, 2017. www.defensenews.com/smr/equipping-the-warfighter/2017/09/28/industrial-base-wargaming-pentagon-wants-companies-to-find-supply-chain-weaknesses/

Mosher, Dave and Samantha Lee. "More Than 14,000 Hunks of Dangerous Space Junk are Hurtling Around Earth – Here's Who Put It All Up There." *Business Insider*. March 29, 2018. www.businessinsider.com/space-junk-debris-amount-statistics-countries-2018-3

"Navy Missile hits Dying Spy Satellite, says Pentagon." *CNN*. February 21, 2008. www.cnn.com/2008/TECH/space/02/20/satellite.shootdown/

"New Kids on the Block: How New Start-Up Space Companies Have Influenced the U.S. Supply Chain." Bryce Space and Technology, June 2017. https://brycetech.com/downloads/Start_Up_Space_Supply_Chain_2017.pdf

Office of the Assistant Secretary of Defense for Homeland Defense and Global Security. *Space Domain Mission Assurance: A Resilience Taxonomy*. September 2015. http://policy.defense.gov/Portals/11/Space%20Policy/ResilienceTaxonomyWhitePaperFinal.pdf?ver=2016-12-27-131828-623

"President Signs Space Traffic Management Policy." Department of Commerce, Office of Space Commerce. Press release, June 18, 2018. www.space.commerce.gov/president-signs-space-traffic-management-policy/

Ritche, Ian. "Remote Control Southern Hemisphere SSA Observatory." EOS Space Systems. www.eos-aus.com/wp-content/uploads/2018/08/Advanced-Maui-Optical-Space-Surveillance-Paper-2013-Southern-Hemisphere-Space-Situational-Awareness-Observatory-1.pdf

Shelton, William. "The US Future in Space." The Atlantic Council, Washington, DC, July 23, 2014. www.atlanticcouncil.org/news/transcripts/transcript-the-us-future-in-space

Someswar, G. Manoj., T.P. Surya Chandra Rao, Dhanunjaya Rao. Chigurukota. "Global Navigation Satellite Systems and their Applications." *International Journal of Software and Web Sciences* 12.136 (2013): 17–23. www.unoosa.org/documents/pdf/icg/ISWI/IJSWS12-326.pdf

"Space Fence." Lockheed Martin. Accessed September 6, 2018. www.lockheedmartin.com/en-us/products/space-fence.html

"Space Policy Directive 3 Brings Space Traffic Coordination to Commerce." U.S. Department of Commerce. Press release, June 18, 2018. www.commerce.gov/news/press-releases/2018/06/space-policy-directive-3-brings-space-traffic-coordination-commerce

"Space Situational Awareness." ExoAnalytic Solutions. Accessed September 6, 2018. https://exoanalytic.com/space-situational-awareness/

"Space Surveillance Telescope Australia." M3. Accessed September 6, 2018. http://m3eng.com/portfolio/space-surveillance-telescope-australia-2/

Steinhauser, Lucas and Scott Thon. "Operation Burnt Frost: The Power of Social Networks." *ASK Magazine* 31 (June 1, 2008). https://appel.nasa.gov/2008/06/01/operation-burnt-frost-the-power-of-social-networks/

Strassler, Robert B. *The Landmark Thucydides: A Comprehensive Guide to the Peloponnesian War*. New York: Free Press, 1996.

The White House. *National Space Transportation Policy*. November 21, 2013. https://obamawhitehouse.archives.gov/sites/default/files/microsites/ostp/national_space_transportation_policy_11212013.pdf

The White House. *President Donald J. Trump is Achieving a Safe and Secure Future in Space.* Fact Sheets. June 18, 2018. www.whitehouse.gov/briefings-statements/president-donald-j-trump-achieving-safe-secure-future-space/

The White House. *Remarks by President Trump at a Meeting with the National Space Council and Signing of Space Policy Directive-3.* Remarks. June 18, 2018. www.white house.gov/briefings-statements/remarks-president-trump-meeting-national-space-council-signing-space-policy-directive-3/

The White House. *Remarks by Vice President Pence on the Future of the U.S. Military in Space.* Remarks. August 9, 2018. www.whitehouse.gov/briefings-statements/remarks-vice-president-pence-future-u-s-military-space/

The White House. *Space Policy Directive-3, National Space Traffic Management Policy.* Presidential Memoranda. June 18, 2018. www.whitehouse.gov/presidential-actions/space-policy-directive-3-national-space-traffic-management-policy/

Thucydides. *The Peloponnesian War-The Complete Hobbes Translation.* Edited by David Grene. Chicago: The University of Chicago Press, 1989.

Trench, Marianne. *Cyberpunk* (Documentary). Intercon Production, 1990.

U.S. Department of the Navy. *The Commander's Handbook on the Law of Naval Operations.* NWP 1–14M. July 9, 1995.

United Nations General Assembly. Resolution 2222 (XXI). *Treaty on Principles Governing the Activities of States in the Exploration and Use of Outer Space, including the Moon and Other Celestial Bodies,* or *The Outer Space Treaty.* 1967. www.unoosa.org/oosa/en/ourwork/spacelaw/treaties/outerspacetreaty.html

United Nations. *Charter of the United Nations and Statue of the International Court of Justice.* San Francisco, June 26, 1945.

U.S. Commission to Assess United States National Security, Space Management and Organization, also known as the *Space Commission Report.* January 11, 2001.

U.S. Government Orbital Debris Mitigation Standard Practices. 1997. www.iadc-online.org/References/Docu/USG_OD_Standard_Practices.pdf

Vedda, James A. and Peter L. Hays. "Major Policy Issues in Evolving Global Space Operations." The Mitchell Institute of Aerospace Studies, February 2018. www.aerospace.org/publications/policy-papers/major-policy-issues-in-evolving-global-space-operations/

Vedda, James. "Orbital Debris Remediation Through International Engagement." Crowded Space Series Paper #1. The Aerospace Corporation, March 2017. http://aerospace.wpengine.netdna-cdn.com/wp-content/uploads/2017/09/DebrisRemediation.pdf

Weeden, Brian. "Why Outer Space Matters: Brian Weeden on Natural and Human Generated-Threats on Satellites." *Intercross Blog.* October 24, 2016. http://intercross blog.icrc.org/blog/why-outer-space-matters-brian-weeden-on-natural-and-human-generated-threats-on-satellites

Werner, Debra. "Lockheed Martin Prepares to Turn on U.S. Air Force Space Fence on Kwajalein Atoll." *Spacenews.* May 3, 2018. https://spacenews.com/lockheed-martin-prepares-to-turn-on-u-s-air-force-space-fence-on-kwajalein-atoll/

Wylie, J.C. *Military Strategy: A General Theory of Power Control.* With introduction by John B. Hattendorf. New Brunswick, NJ: Rutgers University Press, 1967; reprint, Annapolis, MD: Naval Institute Press, 1989.

Index